Advance Praise for

Solar Water Heating

Solar Water Heating answers the what? why? and how? of solar thermal water heating system design and installation. Employing down-to-earth language and clear illustrations, this book succeeds in demystifying solar thermal in a way no other book has done. I take great pleasure in recommending this book to everyone I can. Information is power and this book is loaded with it.

— MIKE HELFMAN, president, Bubbling Springs Solar

Bob's book will be required reading for all of our new installation personnel since it provides a complete and comprehensive history, guide and information about the type of work we do every day. It is nice for our new people to get a solid overview of the many types of technologies and strategies for using solar hot water.

— RICHARD LANE, Managing Partner, Solar Mining Company

Bob Ramlow's *Solar Water Heating* is a practical solar encyclopedia for anyone interested in installing a solar energy system or in starting a solar energy company. The folksy writing is clear and the numerous diagrams make this book a great resource for everyone from the novice just getting started to the existing solar professional looking to pick up some of Bob Ramlow's many useful observations from a long career installing and maintaining solar energy systems.

— DON WICHERT, Director, Renewable Energy Programs,
Wisconsin Energy Conservation Corporation

It's high time someone wrote a comprehensive but accessible overview of solar heating, which is one of the most cost-effective steps that we can take to wean from fossil fuels. Bob Ramlow is a rare combination of plumber, engineer, and public educator who has learned with his own hands. Whether you wish to do it yourself or simply be an informed consumer, start with *Solar Water Heating*!

— WINDY DANKOFF, solar industry pioneer and educator since 1975,
and founder of Dankoff Solar Products

SOLAR
WATER HEATING

A COMPREHENSIVE GUIDE TO SOLAR WATER AND SPACE HEATING SYSTEMS

REVISED & EXPANDED EDITION

BOB RAMLOW and BENJAMIN NUSZ

NEW SOCIETY PUBLISHERS

Cover design by Diane McIntosh.
Cover photo © Edward Caldwell. All rights reserved.
Interior illustrations by Benjamin Nusz, unless otherwise indicated.

Printed in Canada by Friesens. First printing April 2010.

Paperback ISBN: 978-0-86571-668-1

Inquiries regarding requests to reprint all or part of *Solar Water Heating* should be addressed to New Society Publishers at the address below.
To order directly from the publishers, please call toll-free (North America) 1-800-567-6772, or order online at www.newsociety.com

Any other inquiries can be directed by mail to:

New Society Publishers
P.O. Box 189, Gabriola Island, BC V0R 1X0, Canada
(250) 247-9737

New Society Publishers' mission is to publish books that contribute in fundamental ways to building an ecologically sustainable and just society, and to do so with the least possible impact on the environment, in a manner that models this vision. We are committed to doing this not just through education, but through action. This book is one step toward ending global deforestation and climate change. It is printed on Forest Stewardship Council-certified acid-free paper that is **100% post-consumer recycled** (100% old growth forest-free), processed chlorine free, and printed with vegetable-based, low-VOC inks, with covers produced using FSC-certified stock. New Society also works to reduce its carbon footprint, and purchases carbon offsets based on an annual audit to ensure a carbon neutral footprint. For further information, or to browse our full list of books and purchase securely, visit our website at: **www.newsociety.com.**

Library and Archives Canada Cataloguing in Publication

Ramlow, Bob
Solar water heating: a comprehensive guide to solar water and space heating systems / Bob Ramlow and Benjamin Nusz. — Rev. and expanded ed.
Includes index.

ISBN 978-0-86571-668-1
1. Solar water heaters. 2. Solar space heating. 3. Solar heating. 4. Solar collectors. I. Nusz, Benjamin II. Title.

TH7413.R34 2010 697'.78 C2010-902533-4

NEW SOCIETY PUBLISHERS

CONTENTS

Books for Wiser Living recommended by *Mother Earth News*

TODAY, MORE THAN EVER BEFORE, our society is seeking ways to live more conscientiously. To help bring you the very best inspiration and information about greener, more sustainable lifestyles, *Mother Earth News* is recommending select New Society Publishers' books to its readers. For more than 30 years, *Mother Earth* has been North America's "Original Guide to Living Wisely," creating books and magazines for people with a passion for self-reliance and a desire to live in harmony with nature. Across the countryside and in our cities, New Society Publishers and *Mother Earth* are leading the way to a wiser, more sustainable world.

Acknowledgments

MOST AUTHORS PAY TRIBUTE TO THEIR FAMILIES, and for good reason. My family have supported me over the years, have encouraged me to fulfill my desire to pursue a socially responsible career and have all contributed to this project. I pay special tribute to my wife, Marguerite, who is my companion on this path.

It was Dr. George Becker who showed me his solar heating system back in 1971. A light went on in my head that day that has not been extinguished. It was Stephen Morris who encouraged me to start writing about solar power and planted the seeds for this book many years ago. And it was the good judgment of my daughter Chamomile to marry Benjamin Nusz, who has diligently worked with me on this project. His enthusiastic help has made this project happen.

Lastly, I would like to acknowledge the plumbers and solar installers who taught me long ago that there is no substitute for quality. Without the groundwork they laid, I would not be here today, putting the finishing touches on this text.

For the Earth,
Bob Ramlow

I consider myself extremely fortunate to have had the pleasure to work alongside Bob. He has been a constant mentor, and I must first thank him for his generosity and dedication. It is a rare opportunity to be able to partner with someone with so much genuine knowledge, and it is one that I am glad to not have missed.

I am especially grateful for the support provided by my family. To my love, Chamomile, and to my sons, Miles and Xavi, I am lucky to have you.

I would like to acknowledge our editor, Audrey Dorsch, and everyone at New Society, particularly Ingrid, Sue, and Jean. Thank you for your persistence and patience.

Finally, I must recognize in advance the hard work that is yet to be put forth by the next generation of solar installers. As I wrote, I had you in mind all along. I hope you return the favor by installing many, many successful solar water heating systems.

Sunny regards,
Benjamin Nusz

INTRODUCTION

EVERY DAY, WE ARE BARRAGED WITH INFORMATION about global warming, global pollution, wars over energy resources, species depletion ... the list goes on and on. Tragically, most people — and certainly most governments — are doing very little about it.

This book tells some of the things we can do. First it talks about using technologies we already have to lower our energy consumption. Then it details renewable energy options for replacing fossil fuels and nuclear energy sources. It shows that as individuals we can change the world!

Most North Americans don't think about our energy consumption very much. We don't think about the energy used to create our society or the energy required to produce the products we consume. We get our heating fuels automatically delivered to our homes. Our electrical energy flows to us through silent wires that are everywhere. It's all very easy and painless. Sure, we think about our energy consumption when we pay our bills once a month, or we think about it when we refuel our vehicles at the gas station, but in most cases we just accept these costs as inevitable. Our society uses a lot of energy; we want as much as we can get; and we want it cheap.

The United States consumes more energy per capita than any other country in the world. Other developed countries with similar lifestyles, and with equal gross national products, consume up to 50 percent less energy. We obviously have the technology today to use a lot less energy than we do. Some experts say that we can reduce our consumption by up to 75 percent today. They surmise that if we would put our greatest minds to the problem, we could do even more.

Lowering our energy consumption and using renewable energy would accomplish several things. Decreased use of fossil fuels and nuclear energy would have a positive impact on our environment by reducing carbon emissions that contribute to global climate change. It would also reduce the production of pollution that contributes to acid rain, nuclear waste and other environmental contaminants. A more subtle result of these changes is that power would be transferred from the superrich to the common people. The people who run the world are the people with the most money. Many of them get their money by selling fossil fuels. These are the people who control the governments and the mass media of the world. By reducing their power we decrease their influence and increase our personal freedom.

It is unlikely that we can look to governments for leadership in the transition to renewable energy that must be made. The changes will come from the bottom up. It is up to us to do what we can.

This book presents detailed information about solar thermal applications for homes and small businesses. Heating water with the sun may seem like a simple endeavor, and it is. Solar water heaters heat water using the energy of the sun. A solar water heating system is composed of a solar collector and a water-storage tank. Depending on the type of solar energy system, it may have several other components as well. Most systems use pipes, pumps, system controllers, heat transfer fluids (also called solar fluids), various valves, heat exchangers, pipe insulation and mounting hardware. Exact system components will be detailed in the following chapters as we describe the different systems available.

Of all the renewable energy options open to us, the technology of solar water heating is the most mature. It is an industry with proven technologies, established manufacturing facilities, and qualified and experienced technicians. These technologies and applications have been proven over many years and thousands of installations. However, it took time for these technologies to mature. And during that time, we made some mistakes. Some of the technologies and designs that are included in this book are bad ideas. We have included their descriptions so you can recognize them and know how they work if you are doing repairs. We have also included these descriptions because you may read or hear about a certain design, and you will want to know if it is good or not. Their descriptions should be used only for reference and maintenance.

It does not take a rocket scientist to design, install or use a solar water heater. Although it is simple, certain factors must be taken into consideration to ensure satisfactory reliability and performance. The single most important consideration is your climate. Others include the hardness of your water; hot water demands (the load); aesthetics; specific location factors such as trees, shading and mounting options; availability of components; and price. With more than 40 years of combined experience in designing, selling, installing and servicing solar water heaters, we have found that most systems work great for a very long time. Problems that do occur are generally caused by poor-quality components, by the wrong design or wrong components for the climate or specific situation, or by poor workmanship. The bottom line is that if you choose the right system for your climate and install it properly, using quality components, you will end up with a renewable energy system that will most likely last you for the rest of your life. It will end up being one of the best investments you ever make.

The first chapter, a short history of solar energy, puts into perspective how we got to where we are today. As you read about the history of solar energy technologies in Chapter 1, you will see that the largest application of these technologies in the world has been in solar water heating.

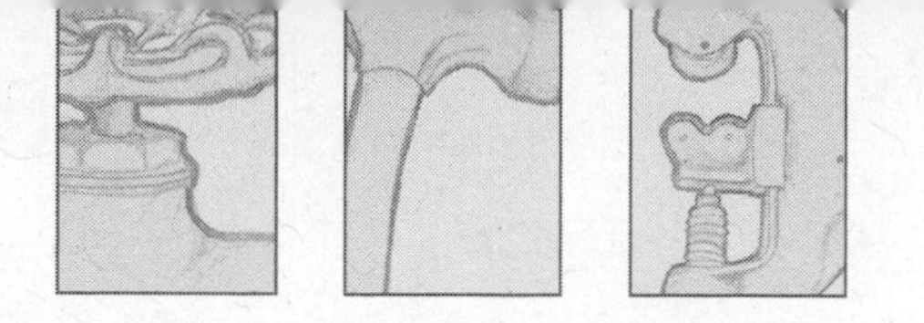

1

HISTORY

The sun is the center of our solar system. The energy it releases warms our planet and powers all life on earth. Through photosynthesis, solar energy is transformed into organic matter — the food that makes our life possible. The fossil fuels we use are actually stored solar energy. Solar energy is also incredibly abundant. Half a day's sunlight falling on the US provides enough energy to run our country for one year.

We often think that modern societies were the first to use solar energy. Not true. Early cave dwellers preferred caves that had openings facing southeast. This allowed the morning sun to warm them up without overheating in the warm months. Native Americans in the Southwest oriented their pueblo dwellings so the low winter sun would heat the buildings by direct solar radiation. Cliffs and overhangs blocked the sun during the summer months, helping to keep the dwellings cooler when the sun was high in the sky.

The ancient Greeks, with a climate that was sunny almost year round, built their houses to take advantage of the sun's rays during the moderately cool winters and to avoid the sun's heat during the summer. Modern excavations of many classic Greek cities show that individual homes were oriented toward the south and entire cities were planned to allow equal access to the winter sun. It is interesting to note that by 500 bc, when the Greeks had almost completely deforested their whole country and needed to find a reliable alternative fuel source, they chose solar energy.

The Roman Empire advanced solar technology by adapting home-building design to various climates, using clear window coverings, such as glass, to enhance the effectiveness of solar heating and expanding solar architecture to include greenhouses and huge public bath houses. Solar architecture became so much a part of Roman life that sunrights guarantees were eventually enacted into Roman law. This society depleted its forest resource as well.

After the fall of the Roman Empire, the use of glass to enhance solar gain in buildings was mostly forgotten. Interest in passive solar architecture and greenhouses was rekindled during the Renaissance. As technologies advanced, glass manufacturing was revived, resulting in an increased use of glass windows. This also made large greenhouses possible for agricultural purposes as well as for recreation.

In the 1700s, a leading naturalist named Horace de Saussure began to experiment with solar hot boxes. These precursors to today's active solar collectors were simple insulated boxes painted black on the inside and with one side made of glass. They were very similar to today's solar cookers and, in fact, many early experimenters used their hot boxes for cooking. Many of the solar principles we use today were identified during those early experiments. Unfortunately, these experiments resulted in few successful applications.

During the late 1800s, domestic water piped directly into homes became more common. Like today, this water supply was cold. People soon wanted hot running water. At first, all water heaters were either coal or wood fired. In 1891 Clarence M. Kemp patented the world's first commercial solar water heater, called the "Climax." It was a black-painted water tank mounted in an insulated box with one side made of glass.

The Climax was instantly popular in California, where it could be used year round. Thousands of Climaxes and similar systems were installed in a short time. They all fall into what we now call "batch-type" solar water heaters: the sun heats the water directly in the tank(s) and the hot water is stored right in the collector tank(s).

In 1909 a California engineer named William J. Bailey began selling a new system he called the "Day and Night" solar water heater. It consisted of a solar collector and a separate storage tank mounted above the collector. His tanks were among the first to be insulated for better heat retention, and his collectors consisted of a pipe grid attached to a flat plate and enclosed in a compact, glazed and insulated enclosure. Cold water dropped into the collector, where it was heated by the sun. As the water was heated, it rose into the insulated storage tank for later use. Today, we call these heaters "flat plate" collectors.

In 1913 a freak cold snap hit southern California and many Day and Night collectors froze and burst. To eliminate future freezing problems, Bailey installed a coil of pipe within the storage tank to act as a heat exchanger. Then he used an alcohol and water mixture as the antifreeze solution for his heat exchange medium. As the sun warmed the solar fluid (or heat transfer fluid), it rose to the storage tank heat exchanger. As the heat from the solar fluid was transferred to the water in the storage tank, the solar fluid cooled and dropped back to the collectors for further heating. This system is described today as a "closed-loop" solar water heating system.

Between 1920 and 1930, huge deposits of natural gas were found in the Los Angeles area. To capitalize on this new, cheap fuel source, Bailey began to manufacture a thermostatically controlled gas water heater. Sales of his gas water heater took off, and sales of solar water heaters plummeted. Gas companies offered generous incentives to hook up to their new gas lines, further hindering sales of solar heaters. Bailey made his last batch of solar water heaters in 1941.

During this same time period, entrepreneurs took the California solar water heater designs to Florida and met with great success. In a building boom between 1935 and 1941, up to 60,000 systems were installed. More than half the population of Miami used solar water heaters by 1941, and 80 percent of the homes built between 1937 and 1941 were solar equipped.

World War II all but halted solar water heater installations. Copper was a major component of solar water heaters, and the use of copper was frozen for all nonmilitary use. When the war was over, solar companies came back, but other factors soon led to their decline. Existing solar water heaters were too small to meet the new, increased demand for automatic washing machines, automatic dishwashers, and other similar appliances. In a final blow, electrical rates fell to half the cost they had been before the war, making electric water heating much more affordable. In an aggressive campaign to increase electrical consumption, Florida Power and Light even offered free installation of electric water heaters. By this time, many of the original, aging solar water heaters were experiencing leaking tanks and plugged pipes. Many homeowners found it cheaper to install an inexpensive electric water heater than to fix their solar water heating systems.

In the United States, the 1950s and '60s were years of unbridled energy consumption. For all but a few people, solar energy was a nonissue. This changed with the first Arab oil embargo in 1973, when Americans experienced long lines at gas stations, limited supplies of other oil products, such as heating fuel, and energy prices that doubled and tripled. President Jimmy Carter helped make energy efficiency and the use of renewable energy a national priority, symbolized by his donning a sweater and installing a solar water heater on the White House roof.

The oil embargo profoundly changed the United States. Coming at the end of the Vietnam War, it added to America's realization of its vulnerability. For the first time since World War II, Americans looked at the way they used energy. Consumers began to demand higher energy-efficiency standards in everything from homes to automobiles. People also looked to renewable energy sources to replace some of the fossil fuels they were using.

The whole nation took on the challenge of reducing its dependence on oil from the Middle East. Renewable energy sources were rediscovered, and new companies sprouted everywhere to fill the growing demand. Government spending on renewable energy research and development increased from about $1 million to more than $400 million. While this was a small fraction of the attention and money given to the nuclear industry, it was a dramatic change nonetheless.

During the late '70s and early '80s, installing solar energy systems was seen as patriotic. The federal government, as well as many state governments, passed legislation encouraging the use of solar energy systems through tax credits. Federal incentives combined with state incentives (where available) often offset more than 50 percent of the cost of many renewable energy systems. A new renewable energy boom began. People looked to wind-powered electric systems, active space heating systems, advanced passive solar-heating systems, the newly emerging solar electric systems, and advances in energy-saving technologies as well as the old reliable solar water heaters.

Most of the solar energy companies that sprang up in the 1980s were reliable firms that installed quality systems. Unfortunately, with the general public's headlong plunge into the use of renewables, a few companies selling inferior products and doing inferior work joined the fray. Some brought products to the market without proper testing. Others just wanted to make a quick buck and didn't care if they were taking advantage of well-intentioned consumers. Although most renewable-energy systems were of good quality, the minority that weren't gave solar a bad name.

The young solar industry was experiencing the typical growing pains that come with most emerging technologies and took steps to correct the problems. The federal government, as well as many state governments, also stepped in to ensure higher quality.

This move toward renewable energy did not sit well with those who profited from selling fossil fuels. After the most expensive presidential campaign ever, financed in part by oil interests, Ronald Reagan became president of the United States. His presidency heralded a return to fossil fuels. One of his first acts as president was to remove the solar water heater that President Carter had installed on the White House. Between 1981 and 1986, Reagan effectively gutted the US solar industry. He negotiated a repeal of the tax credit legislation for renewables that was in effect. He reduced funding for renewable energy by 90 percent. He also spearheaded a massive campaign to discredit renewable energy. The result was a 91 percent drop in the sales of solar hot water collectors between 1984 and 1986. The solar market in North America from the 1980s through the late 1990s was primarily supported by customers who wanted to invest in renewable energy for environmental reasons. For an in-depth analysis of this subject, I invite you to read *Who Owns the Sun?* by Daniel Berman and John O'Connor. Their thorough and thought-provoking book will change the way you look at energy and politics forever.

While North America was abandoning its use of renewable energy, virtually all other developed nations in the world continued to embrace renewables and energy efficiency. As a result of this continued effort, our European and Asian trading partners reduced their energy consumption in relation to gross national product by as much as one half that of North Americans. And while maintaining a similar lifestyle, they also reduced the energy need to half that of the United States. Some countries, such as Israel, require that solar water heaters be installed on all new dwellings and businesses. Developing nations are also embracing renewable energy technologies over traditional fossil-fuel energy sources. Although their track record is not perfect, their attitude toward renewable energy is

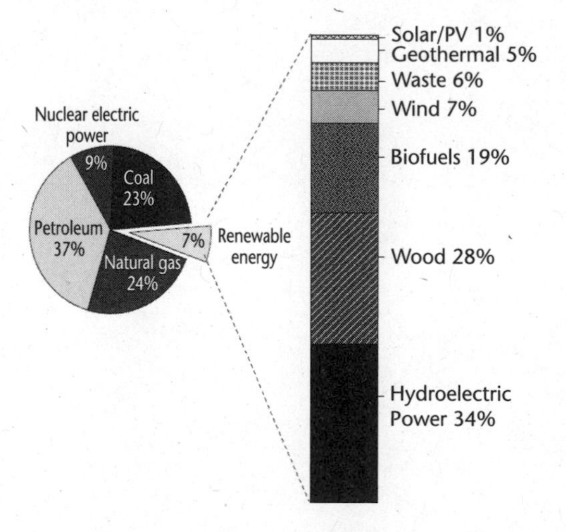

Figure 1.1: *US energy consumption, 2008*

significantly better than that of North Americans.

In Figure 1.1, you can see the breakdown of the total energy consumption for the US in 2008. Renewable energy accounted for only seven percent! Solar, both thermal and electric, made up only one percent of the renewable energy share — only 0.07 percent of the total.

Now that we've covered history and the present, we need to look at our future. If what happened in the past carries on through tomorrow, we are headed down a dark and dirty path. One percent of only seven percent is not good enough and needs to change now. Let's do all we can to make our children's future brighter.

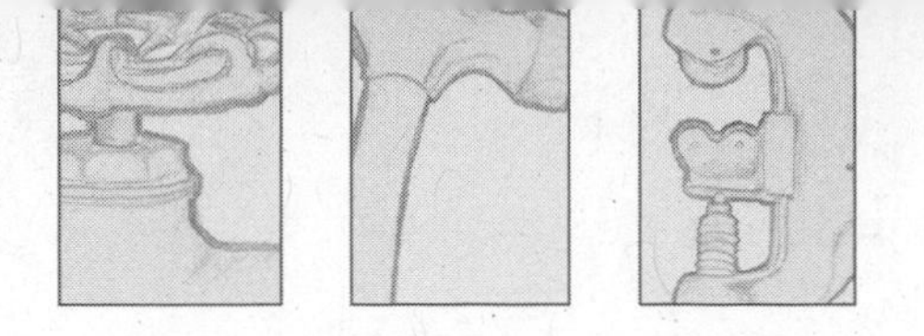

2

CONSERVATION AND THE ECONOMICS OF SOLAR WATER HEATING

WHATEVER BROUGHT YOU to this point, whether it was the realization that we are trashing our environment or the simple need to lower your living expenses, now you are here and you want to do something. But what to do first? The answer is simple: start by conserving the energy you use to heat water. Three general principles that are easy to follow will also save you money: reduce losses, increase efficiency and reduce consumption.

To start, examine your heating system from top to bottom and look for places where heat might leak out. Heat losses in the system end up wasting the energy you just used to heat your water. Many losses can be reduced with just a bit of cheap insulation. For instance, insulate all your hot pipes. If you are working on a new construction, insulating the hot pipes is easy. Even if you don't have access to all your pipes, insulating the ones you can get at will make a noticeable difference. You should also insulate your water heater. A tank type water heater heats a whole batch of water. As this water sits there waiting for use, it slowly cools down. The more you insulate it, the better it will retain its heat. Heat losses can also come from leaks. A faucet that leaks 30 drops of water a minute will waste almost 100 gallons a month. Fix leaky faucets promptly.

Next, try to increase the efficiency of everything in your home that uses hot water, for instance, the washing machine and the dishwasher. Upgrading these appliances to more energy-efficient models will significantly reduce the amount of energy consumed. A frontloading washing machine uses half the hot water of a standard top-loading model. This results in saving 10 to 20 gallons of hot water in each load you do. You can save thousands of gallons of hot water a year.

Finally, you can conserve energy by simply using less. Former US Vice President Dick Cheney is famously quoted as saying that the "American way of life is non-negotiable." We couldn't disagree more. Simple behavior modifications can dramatically reduce the amount of energy we consume without a significant change in daily habits. For instance, when washing dishes in the sink by hand, don't let the water run while rinsing. Fill one sink with wash water and the other with rinse water. Soak pots and pans instead of letting the water run while you scrape them clean, and if you are using a dishwasher, wash only full loads. Use cold water with the garbage disposal. Cold water solidifies grease, allowing the disposal to get rid of it more effectively. You can take short showers instead of baths. You should first install a low-flow shower head. Most standard showerheads use three to four gallons per minute. Even if you take a relatively brief five-minute shower, you can consume 20 gallons of hot water. Low-flow showerheads will use half of that. A family of four can save well over 1,000 gallons a month. If you are particularly attached to your showerhead, you can install a flow restrictor that will reduce the number of gallons per minute that it uses. For only a couple of dollars, you can reduce your load substantially.

Reducing losses, increasing efficiency and reducing consumption: these are the first steps. More important than quick fixes, though, is the notion of conscious consumption. We have forgotten the financial and environmental costs of hot water. If everyone recognized that whenever we turn on the hot water faucet we are using energy produced by nonrenewable sources, this would reduce energy consumption more than any other measure.

People often say to us, "I have done a lot of energy conservation and now I am ready to invest in a renewable energy system. What should I do next?"

Today, homeowners and business people can choose from a wide range of renewable energy technologies. Popular options include photovoltaic (solar electric) systems, wind electric systems and solar water heaters. In almost every case, a solar water heating system is the best place to start. It provides a higher return on your investment than most other types of renewable energy systems. A solar water heater works 12 months a year, providing hot water to your home or business with little or no additional cost, thus offsetting your previous bill for heating water with conventional energy sources. Depending on your particular situation, the savings in conventional fuel can pay for the cost of the solar water heating system in as little as three years. Most often the payback is five to ten years — still a great investment, even without taking into account the ecological benefits of not burning all that fossil fuel.

In fact, since you've already bought this book, it's time to let you in on a little secret. Solar water heaters don't cost anything. They're FREE! Of course, it may sound absurd, but it's true. Now, we're not recommending that you run over to the nearest solar distributor and just take a system. Don't do that. We are just asking you to take a step back and think about solar in a different way. With a little change in perspective, you will see that in the end, solar water heaters have a net cost of zero dollars.

There are two ways to take this in. The first one is easy: when you install a solar water heater you are increasing your home's value. You gain in equity what you spent on the cost of installation. Many solar water heaters have a life span of 30 to 40 years, and sometimes longer. In most cases, the solar collectors will outlast your roof. So if you decide to sell your home, you should get back most of what you paid for the cost of installation. Though there isn't a lot of data on this fact, the general rule we have seen is that for every $100 in annual energy savings you can expect an additional $2,000 in resale value of the property. For instance, if your solar water heater saves you $200 a year, your fair market value should increase by $4,000. Here in Wisconsin, and in many other states, renewable energy systems are exempt from property taxes. Like any other home improvement, this is an investment, but the added value doesn't carry an additional annual cost.

Yet, just because something retains its value over time and you don't have to pay taxes isn't usually reason enough to go out and buy it. The second part of this shift in perspective takes a bit more explanation, but we assure you it is even more convincing.

The True Cost of Fossil Fuel

We'll start by comparing solar with the alternatives. Unless you are reading this to find out how to fix your existing system, you probably heat your water with some type of fossil fuel, such as natural gas, propane or electricity.

When you purchase fossil fuels, you do not pay anywhere near their whole cost. Because our taxes subsidize the oil companies, for instance, the true cost of gas is not reflected in the price we pay at the pump. Let us say it again: oil companies don't pay taxes on all the money they earn, so we must all pay higher taxes to make up for it. It goes without saying that if they paid their fair share of taxes, our tax rates would be lower, and the price we pay for gas would be higher. The same scenario holds true for all other fossil fuels and electricity.

How can this be? First, the fossil fuel companies are among the richest corporations in the world, with tremendous influence in politics. For nearly a century they have manipulated the government into granting them numerous tax breaks and outright payments that are not enjoyed by any other class of corporation. The end result is that they pay little if any tax but significantly influence how our tax dollars are spent. They have managed to get the government to pay for lots of expensive research for their industry.

The costs to the environment of using fossil fuels are also hidden. Burning fossil fuels releases carbon into the atmosphere, leading to global climate changes that will disrupt life as we know it on every corner of the Earth. The costs of dealing with these changes will be astronomical and are directly linked to burning fossil fuels. When we burn fossil fuels, especially coal, we release into the atmosphere chemicals that cause acid rain, polluting our rivers, lakes and soil. Acid rain kills wildlife, trees and vegetation and degrades our buildings, roads and anything else exposed to it. Although we are already paying some of the costs to fix these problems, we are not paying them all. Eventually, someone will have to pay them.

Then there are health-related costs. Whenever we burn any fossil fuel, we release air pollutants that harm our health. Our health insurance costs go up to help pay for the care required by those most affected. Our taxes are increased to help pay for those who cannot afford their own care, and our general health care costs go up for the same reason. Again, we do not pay these costs at the pump or with our utility bills.

Some of our electricity is generated in nuclear power plants. The waste generated by these plants is one of the most toxic substances known to humanity. We have no clue how to safely dispose of it. We can send people to the moon, but we have not figured out how to deal with these incredibly toxic waste products. Undoubtedly, if we do figure out a way to safely dispose of them, it will be incredibly expensive. This cost is not included when we pay our electricity bill. We also invite you homeowners and renters to read the fine print of your insurance policies. Note that if there is ever an accident involving nuclear fuel or waste, your insurance policy does not cover that. Ask the people living around the Three Mile Island nuclear power plant, whose lives were devastated by the nuclear accident there, how they feel. We will guarantee you that they are not happy. Many lost everything.

Assigning a true cost to the use of fossil fuels relates directly to solar water heaters and any other renewable energy system. Admittedly, it costs money to invest in renewable energy equipment. Often, people will look at that cost and say that it is just too much more than using fossil fuels. When you get to this point, please remember the above discussion. How much higher should the costs of fossil fuels really be? Twice as expensive? Three times? Four times? Experts who have spent considerable time researching this issue have calculated that these costs are five times more than the bill we pay at the pump or meter.

While these true costs are not reflected in our bills, we do see that the price we pay tends to increase every year. In order to understand where fossil fuel prices are heading, we first have to understand what has happened in the past. Having a handle on energy price inflation is basic to understanding the economic impacts of investing in solar thermal energy systems.

A Brief History of Fossil Fuels

Soon after oil was first discovered in the mid-1800s there was a glut, keeping its price very low. With the advent of mass-produced automobiles, there was a steady demand for oil, and prices became stable, rising at a rate at or slightly above the inflation rate. During World War II, fossil fuels were diverted to the war effort, so they became hard to get and more expensive. After the war, fuel again became plentiful and relatively inexpensive. During the postwar period, per-capita consumption of fossil fuels skyrocketed.

It is important to note that until the 1970s, almost all the oil used in the United States was produced here. In 1970 the United States reached peak oil production while demand continued to escalate. About 1996, imported oil overtook domestically produced oil for use in US consumption.

The Oil Embargo

Beginning in 1973, political factors caused a shortage of oil in the worldwide market, followed by a global recession. The shortage continued through the early 1980s. The OPEC oil embargo made people think about how they used energy, and energy conservation became common practice. For the first time, people began to talk about running out of oil on a large scale. In fact, though, during this period there was plenty of oil available and in the ground. The oil spigots could have been opened at any time and the crisis would have been over in a day. In fact, this is essentially what happened in the mid-1980s.

Peak Oil and Natural Gas

Today we are facing an oil shortage much different than that of the '70s and '80s. We are entering the era of peak oil. There are many good books on this subject, such as *The Party's Over* by Richard Heinberg, so we will not go into a lot of detail here. But essentially, today we are at a turning point in the history of modern civilization because the production of oil is at its peak; it will never grow larger, as it has in the past. At the same time, worldwide demand for oil is growing faster than at any time in history. As a result, the price of oil will continue to rise while the supply will decrease. It is important to note that Earth's oil supply is not entirely depleted. About one-half of all the oil there ever was is still left in the ground. The reality is that we have reached peak oil production, while demand for oil continues to rise at record levels. For some time there will still be oil to be had, but producing it will become increasingly more expensive.

You are probably wondering why we have been discussing oil at such length. Though oil is rarely used to heat water, the price of oil affects the price of all other forms of energy. When it goes up, they do too. This is especially true for electricity. It takes large amounts of oil to mine coal, the basic feedstock of most of our electrical generating capacity in the US. It takes oil to mine and process uranium to feed our nuclear power plants. Oil is used in the natural gas exploration and distribution industry. Our society is completely and utterly dependent on a constant flow of cheap oil. As we enter the peak oil era, its price will continue to rise, with no end in sight.

The same holds true for natural gas, which is used to heat water. We have now reached peak natural gas production in North America, where all the natural gas used in the United States is produced. We will never be able to produce more natural gas than today, even if demand rises (which it is doing). When the United States reached peak oil production in 1970, this was a significant milestone, but it was not devastating to our economy or culture. The oil companies could inexpensively import oil from other parts of the world. All they had to do was to build inexpensive oil tankers to ship foreign oil to the US. This is not the case for natural gas. It takes very sophisticated and expensive ships to import natural gas from foreign sources. Also, expensive and sophisticated terminals must be built at both the shipping and receiving ports. This infrastructure is not in place and it will take many years and a substantial investment to create it. This will significantly affect the price of natural gas in the future. The bottom line is that the cost of natural gas will also continue to rise, with no end in sight.

Life-Cycle Costing

People often ask, "Why would I consider purchasing a solar water heater that costs several thousand dollars when I can purchase a gas or electric water heater for only several hundred dollars?" The answer lies in the fact that they do not think about life-cycle costing. Life-cycle costing adds the original cost of a piece of equipment to its operating cost over the equipment's lifetime, or at least over a certain amount of time. Using an analysis like life-cycle costing gives an accurate analysis of the real overall cost of a purchase and allows you to make accurate and informed comparisons.

This is why it is important to figure out the energy inflation rate. We know, for example, that natural gas prices will increase rapidly. The question remains: by how much? The energy inflation rate is simply the percentage more you have to pay each year for the same amount of energy. The average energy inflation rate over the past 35 years for natural gas in the residential sector was about 7.5 percent. At the time of this writing, natural gas prices have been flat for a little while. This is primarily a result of a bad economy in which manufacturing has declined, so the use of natural gas has declined too.

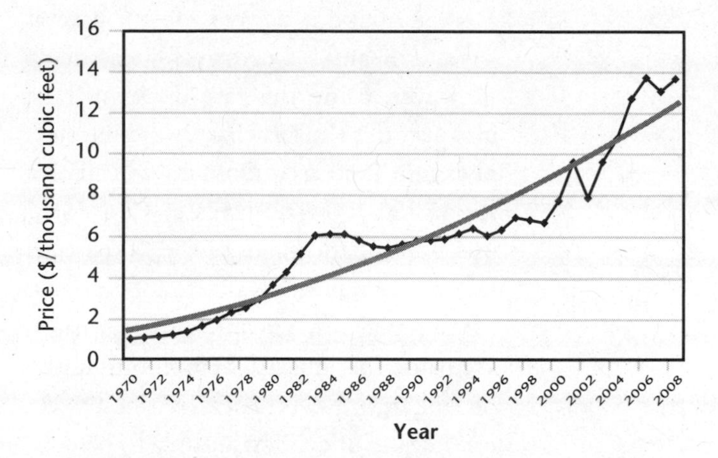

Figure 2.1: *US residential gas price trends*

Do not be fooled into thinking that there is no longer a problem with natural gas supplies. The fact is that the production of new wells is not keeping up with depletion of older wells.

Figure 2.1 graphs the price of natural gas over that time period. You can see the effects of the oil embargo in the late '70s and early '80s and how the surplus led to a rate decrease. You can also see what has been happening lately — the sharpest rate increases since the oil embargo, with inflation rates that surpass those of that era. Over the past five years natural gas has increased at an average rate of 10 percent, and this is just the start.

When calculating energy savings, you can't just use the cost of energy during the first year for each additional year because we know that the cost is rising at such a sharp rate. The lowest energy inflation rate that should be used is the historical average of about 7.5 percent for natural gas and oil and 5.5 percent for electricity. However, using this rate does not take into account the fact that dwindling supplies and increased demand will have a strong impact on rate increases. Conversely, using an inflation rate in the 20 percent range, as we have seen in the last couple of years, may result in unrealistic estimates. We made that mistake before, during the oil embargo. Nevertheless, it is best to err on the side of caution and try to be conservative in your estimates without being unrealistic. When estimating life-cycle costs, we use 10 percent as the energy inflation rate for natural gas and fuel oil, and 7 percent for electricity. We believe that this is the lowest it will realistically be over the next 30 years. Most likely we will see the next 30-year average in the 15- to 20-percent range.

In Figure 2.2 you can see an example of life-cycle costing. This example compares both an electric water heater and a natural gas water heater to a solar water heater. All three systems are producing exactly the same amount of hot water. But solar water heaters have *no operating costs*. You do not have a monthly bill to pay because the solar resource is free. Like any other piece of mechanical equipment, they do require some maintenance, but this amounts to only about $2 per month.

As you can see in the table, viewing the systems in the long term makes for a more fair comparison. The cost of the solar water heater is equal to the operating cost of the electric water heater after only 11 years and the natural gas water heater after 15 years. This number is commonly referred to as the payoff date, because you would have paid for the system with the money saved from not having to purchase energy from the utility. However, we want to stress that this is a misstatement. As we said before, a solar water heater is paid off the second you install it because of what you have gained in equity.

Nevertheless, the notion of a payoff date is still a good way to think about the cost of a solar water heater. For instance, imagine yourself ten years from now. You will have taken just as many showers, washed just as many loads of clothes, essentially used the same amount of hot water, but you had a choice whether to heat this water with polluting coal or with clean energy from the sun. You would have paid the same amount over that time period. In other words, a solar water heater will not cost you a penny more. The bottom line is that over a ten year period the two systems cost essentially the same to purchase and operate. It is in this sense that the solar water heaters are free. Taking cold showers and not washing your clothes isn't a realistic alternative.

After the payoff date, the solar water heater will produce free energy for the duration of the system. It truly is free because you have already offset the cost of installation with money saved. With scheduled maintenance, many solar water heaters will have at least a 40-year life expectancy, so over that time frame the savings from the solar water heating system will be about $190,000. That is a lot of money to save from a one-time investment of $9,000. Because of the cumulative effect of energy inflation, the savings add up quickly, totaling quite an impressive figure. Obviously, it makes economic sense to choose the solar water heater over the electric water heater.

Another way to look at the payoff date is to introduce the idea of pre-payment. Imagine that you were approached by a salesman who gave you a deal. If you pay for the next 10 years' worth of hot water right now, you can get the following 30 years for free. This is what is being offered by a solar water heating system. Up until the pay-off date you are simply buying the hot water from a renewable resource. You are going to use hot water in any case, so why not make the responsible choice? Choose solar.

At the time of writing, a federal tax credit for solar water heaters pays for 30 percent of the cost of an installed system. Many states and utilities also offer rebates for solar. These rebates can drive down the initial installation cost of the systems and will consequently reduce the payoff time. Figure 2.2 also shows the life-cycle cost for a system that was eligible for federal and state incentives. In this case the payoff date is only six years away when compared to an electric water heater and nine years when compared to natural gas. Your eligibility for the various financial opportunities may have a significant impact when making a life-cycle analysis.

Deciding on what fuel inflation rate to use will also affect a comparison like this. If we had a crystal ball that would accurately predict future fuel prices, we would obviously have a more accurate prediction of the actual outcome of the comparison. If energy inflation rates rise at more than just the ten percent used above, the payoff time will be reduced. If the past couple of years are indicative of the future energy inflation rate and it turns out to be 20 percent, the system would be paid off in 7 years. Conversely, if the inflation rate dropped down to the historical average of 7.5 percent, it would take 10 years to pay off the system.

In addition, the life-cycle example does not include the cost of borrowing if the solar energy system or the electric or natural gas water heater has to be financed. Many of us don't have $9,000 readily available to spend on a system that heats

		Electric Water Heater	Natural Gas Water Heater	Solar Water Heater	Solar Water Heater with rebates
Annual Energy Produced		4,300 kWh	200 therms	200 therms or 4300 kWh	200 therms or 4300 kWh
Cost per Unit		$0.11	$1.25	$0.00	$0.00
Energy Inflation Rate		6%	10%		
Installed Cost		$1,500.00	$1,500.00	$9,000.00	$5,000.00
Maintanence		$0.00	$0.00	$2.00/month	$2.00/month
Cost to Operate	1st year	$473.00	$250.00	$0.00	$0.00
	2nd year	$501.38	$275.00	$0.00	$0.00
	3rd year	$531.46	$302.50	$0.00	$0.00
	4th year	$563.35	$332.75	$0.00	$0.00
	5th year	$597.15	$366.03	$0.00	$0.00
	6th year	$632.98	$402.63	$0.00	$0.00
	7th year	$670.96	$442.89	$0.00	$0.00
	8th year	$711.22	$487.18	$0.00	$0.00
	9th year	$753.89	$535.90	$0.00	$0.00
	10th year	$799.12	$589.49	$0.00	$0.00
Total Cost of Energy Consumed		$6,234.52	$3,984.36	$0.00	$0.00
Life Cycle Cost After Ten Years		$7,734.52	$5,484.36	$9,240.00	$5,240.00
After 15 Years		$15,162.17	$7,854.50	$9,360.00	$5,360.00
After 20 Years		$26,128.25	$12,955.00	$9,480.00	$5,480.00
After 30 Years		$72,232.43	$34,598.80	$9,720.00	$5,720.00
After 40 Years		$191,814.80	$90,018.51	$9,960.00	$5,960.00

Figure 2.2: *Life-cycle costing comparison*

water. That means you are off to the bank to take out a loan. The added interest will add some time to the payoff date, but only about a year.

And remember, the price of energy shown on your monthly utility bill overlooks the real cost of burning fossil fuels. Every year it is in use, the solar water heater used in the comparison above would eliminate one to two tons of greenhouse gases for an electric water heater and one ton of greenhouse gases for a natural gas water heater. If the environment had a dollar value, what would it be?

Cash-Flow Analysis

Another way to look at the economics of solar water heaters is to look at a cash-flow analysis. This looks at the impact an investment will make on your cash flow. We all have some method of making money, and we get a certain amount of money regularly as income. Then we spend this income to get the things we need and want. These are our expenses. A sound cash flow is a balance where our expenses do not exceed our income.

A solar water heating investment is different from most investments because the value of the free energy that a solar water heater harvests reduces a bill you would otherwise pay each month. If you heat your water with fossil fuel, you have a hot water bill each month that is part of your normal cash flow. When you install a solar water heater, your hot water bill is reduced. The savings gained from the solar water heater pays for the solar investment.

Figure 2.3 gives an example of a cash-flow analysis for a solar water heater that was used to offset an electric water heater. We used the same starting costs as in the life-cycle analysis but factored in some available rebates. Assume that you need to borrow the entire $5,000 to pay for the solar water heating system at 6 percent interest and make equal monthly payments for ten years. As you can see, the monthly loan payments initially exceed what is saved from not having to purchase electricity. However, over time the two columns level out, making much more comparable figures. Eventually the monthly loan payments are less than the monthly utility bill, meaning that you will actually have more money in your pocket from month to month. You would see a small negative impact on your cash flow for the first six years, and a positive cash flow impact thereafter. After ten years, when the loan has been completely paid off, your cash flow per month is greatly increased

	Electric Water Heater Monthly Savings	Solar Water Heater Monthly Payment	Cash Flow Impact per Month
Monthly Bill			
1st year	$39.42	$55.51	-$16.09
2nd year	$41.78	$55.51	-$13.73
3rd year	$44.29	$55.51	-$11.22
4th year	$46.95	$55.51	-$8.56
5th year	$49.76	$55.51	-$5.75
6th year	$52.75	$55.51	-$2.76
7th year	$55.91	$55.51	$0.40
8th year	$59.27	$55.51	$3.76
9th year	$62.82	$55.51	$7.31
10th year	$66.59	$55.51	$11.08
11th year	$70.59	$0.00	$70.59

Figure 2.3: *Cash flow analysis — solar vs. electric*

Figure 2.4: *Cash flow analysis — solar vs. natural gas*

	Natural Gas Water Heater Monthly Savings	Solar Water Heater Monthly Payment	Cash Flow Impact per Month
Monthly Bill			
1st year	$20.83	$55.51	-$34.68
2nd year	$22.91	$55.51	-$32.60
3rd year	$25.20	$55.51	-$30.31
4th year	$27.72	$55.51	-$27.79
5th year	$30.50	$55.51	-$25.01
6th year	$33.55	$55.51	-$21.96
7th year	$36.90	$55.51	-$18.61
8th year	$40.59	$55.51	-$14.92
9th year	$44.65	$55.51	-$10.86
10th year	$49.12	$55.51	-$6.39
11th year	$54.03	$0.00	$54.03

and will continue to increase as energy prices rise.

Figure 2.4 demonstrates a comparison between solar and natural gas. The cash-flow impact is slightly greater because natural gas tends to be less expensive than electricity. Nonetheless, the additional monthly cost diminishes over time, and after the loan is paid off in ten years all the savings contribute to increasing positive cash flows. Since the system is expected to last about 40 years, you can plan on seeing many years when you will have more money to spend on a monthly basis. When you finance the system, you are essentially locking in your monthly payments. You know what you will have to pay each month and will not be affected by the continually rising cost of energy. Some view this as a retirement investment. They pay off the system now when they have the cash flow to do so, and when they retire, their utility bills and monthly expenses will be greatly reduced.

The point of this analysis is to show you that you should not get hung up on the upfront cost of a solar heating system because the investment does not significantly impact your cash flow. You could have a solar water heater today for only an additional $20 a month.

At the start of this chapter we told you that we would demonstrate how you can get a solar water heater for free. The point of the life-cycle costing and cash-flow examples is to show that no matter how you look at it, a solar water heater will not cost you any more than its alternative. All you have to do is install the system and you can start saving today.

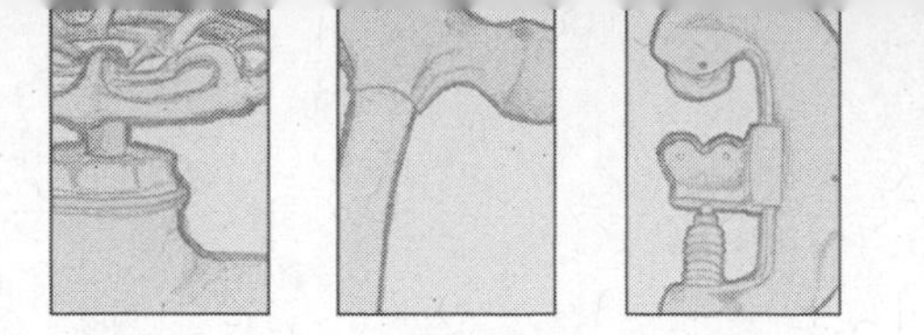

3

TYPES OF SOLAR COLLECTORS

FROM THIS POINT ON, this book will deal with the nuts and bolts of solar water heating and space heating systems. We think it is appropriate to first let you know our perspective. We both live in Wisconsin, where the weather can be severe, particularly in regard to cold temperatures and snow that stays for months. Our experiences with solar thermal systems have been heavily influenced by the weather here. It is interesting to note that approximately 90 percent of the population of the world lives in a warmer climate than we do. Green Bay, Wisconsin, just 50 miles away from our homes and at the same latitude, is the coldest city in the continental US with a population over 100,000. The average coldest temperature in Wisconsin is –44°F. (The record is –56°F.) In a cold climate like this we have seen everything go wrong with a solar thermal system than can possibly happen because of cold weather. We also get a lot of snow each winter, and this snow typically stays on the ground and on roofs for at least four months without melting. Also know that it gets plenty hot here during the summer, so we have seen that side of things as well. The bottom line is that there is no room for error around here in either system design or installation, and this has influenced our perspective significantly.

So we are picky when it comes to system designs. If you follow our suggestions that are based on "worst-case scenarios" you will have success, even if you are in a warmer or less severe climate than ours. The next question is, are we suggesting designs that are overkill? The answer is no, because it does not actually cost more to design and install a system properly whether you are in a hot, cold or in-between type of climate. Having lived through the oil embargo years and the ensuing solar boom, we have seen many systems installed that were not appropriate for the local climate or were sub-standard in other ways. These systems required significant repair costs that, when added to the original cost of the system, resulted in a more costly system than systems that were properly designed and installed in the first place.

Flat Plate Collectors

Flat plate collectors are the most widely used kind of collector in the world for domestic solar water heating and solar space heating applications. These collectors have an operating range from well below 0°F to about 180°F, which is precisely the operating range required for these applications. They are durable and effective. They are the standard to which all other kinds of collectors are compared.

Figure 3.1: *Flat plate collector*

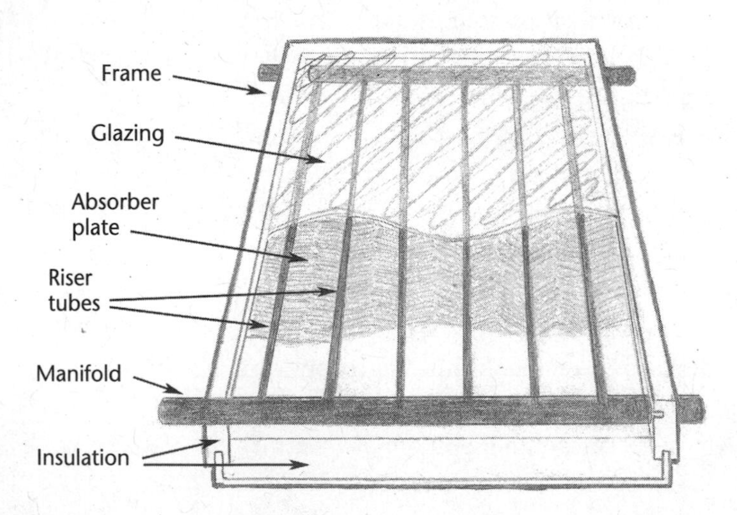

Flat plate collectors are rectangular shallow boxes that typically are 4 feet wide, 8 or 10 feet long and 4 to 6 inches deep, but they also come in other sizes. A common metric size is 1 meter by 2 meters. These collectors are made with a strong frame, a glazed (glass) front, a solid back and insulation on the sides and back. An absorber plate lies just beneath the glazing. In most cases this absorber plate has manifolds that run across the top and bottom of the collector, just inside the frame. These manifolds are usually ¾-inch or 1-inch-diameter copper pipe and extend out both sides of the collector through large rubber grommets.

These collectors, called internally manifolded collectors, can be easily ganged together to make large arrays. Smaller riser tubes, typically ½-inch copper pipes, run vertically, usually brazed to the manifolds above and below, and are spaced 3 inches to 6 inches apart (the closer the better). Another type of absorber plate has a serpentine tube that meanders back and forth continuously from the bottom of the collector to the top of the collector.

A flat copper or aluminum fin is then attached to each riser to complete the absorber plate. The fin must make intimate contact with the riser tube to facilitate effective heat transfer from the fin to the tube. Soldering, welding or roll bonding the fins to the tubes makes the best connection. The plates are also usually dimpled or corrugated to increase absorptivity. The absorber plates are not attached to the frame; they just sit inside it and can expand or contract as they are heated or cooled without being restricted by the frame.

Collector Frame

It is very important that a flat plate collector have a strong frame. The collector mounting hardware is fastened to the frame, and strength is very important because the collectors must be able to withstand high wind conditions without breaking. These frames are almost exclusively made of extruded aluminum, although some are made of rolled aluminum or extruded fiberglass. The heaviest-duty collectors use heavy, thick, extruded aluminum frames. The extruded aluminum frames have channels, or flanges, built into them that the mounting hardware fastens to. Because these flanges go completely around the collector, great flexibility in mounting options is available.

Another important component to look at when considering a collector is the kind of fasteners used to assemble the collector. All fasteners should be made of stainless steel. It is critical to use compatible metals where they are attached to each other. Aluminum and stainless steel are compatible; aluminum and plain or galvanized steel are not. This must be applied not only to the construction of the collector but also to the mounting hardware. Because each manufacturer makes its own mounting hardware, and because each collector is tested with its specific hardware, you should always purchase your mounting hardware to match your collectors.

Glazing

All kinds of plastics have been used as glazing material for collectors, but they have all failed under direct, constant exposure to the sun. Only low iron tempered glass has stood the test of time. Iron in glass causes some of the solar radiation to be absorbed by the glass, diminishing the solar radiation hitting the absorber plate. The glass is usually patterned on one side to reduce glare and reflection. A rubber gasket is fitted to the edges of the glass plate both to protect the edge and to create a good seal where it sits against the collector frame. Some collectors use a silicon caulk to seat the glass against the frame. Though this method does last, it makes it almost impossible to remove the glass when making repairs.

Note that if you ever have to take the glazing off a collector, the edge of the tempered glass is very fragile. If you even tap the edge or side of a tempered-glass pane, it can literally explode, so be very careful and always wear safety glasses and gloves when handling glass.

Some early collector models used either a double pane of glass or a thermopane in an attempt to minimize heat losses from the front of the collector. However, over time we have learned that the second sheet of glazing actually lowered collector efficiency because it reduced the amount of solar radiation that could reach the absorber plate. To our knowledge, no flat plate collectors are made this way any more, but you may still encounter them in service calls.

Insulation

To minimize heat losses, flat plate collectors have insulation on the back and sides of the collector. Common insulation types are polyisocyanurate, rigid expanded polyurethane (PUR) and mineral wool (fiberglass, rock wool). All are commonly used and are capable of withstanding the prolonged high temperatures experienced inside the collector.

Absorber Coating

Absorber plates are necessary to conduct the solar radiation to solar fluid. How the absorber has been coated will directly affect the efficiency of this process. Absorber coatings are rated along two parameters: absorptance and emittance. The former refers to the percentage of solar radiation that can be absorbed, and the latter is the percentage of heat that is emitted back from the absorber plate. To get the net heat gain, you need to subtract what is emitted from what is absorbed. Most coatings will have similar absorptance ratings in the range of 90 to 98 percent, but will vary in their levels of emittance. Traditional flat or selective black paints will emit anywhere from 15 to 30 percent of their heat, and the modern high-tech methods, such as sputtering, physical vapor deposition (PVD), black chrome or black crystal, can reduce the emittance levels to 5 to 10 percent.

Evacuated Tube Collectors

When you look at a flat plate collector you will see that the sides and back are well insulated. However, there is no way to insulate that large pane of glass because it would block all the solar radiation. This obvious limitation has led to the development of evacuated tube collectors. Inventors sought a way to permit the transmittance of solar radiation while still insulating.

Evacuated tube collectors are constructed of a series of glass tubes. Each tube is made of annealed borosilicate (Pyrex) glass and has an absorber plate within the tube. During the manufacturing process, a vacuum is created inside the glass tube. The absence of air in the tube creates excellent insulation, allowing higher temperatures to be achieved at the absorber plate by minimizing heat losses. Air is the medium in which convective heat is transferred. If all the air is removed from the tube, this method of heat movement is interrupted.

Because evacuated tube collectors are able to minimize heat losses they are able to achieve higher temperatures than other collector types. This can be an advantage or a disadvantage depending on how the system has been designed and/or what climate the collector is installed in. For applications that require high temperatures, such as solar cooling, evacuated tubes are the most common choice. However, care must be taken when planning and installing these collectors to ensure that the fluid does not overheat and boil during periods of stagnation or when the load on the system is low. Whenever the sun is out, you need to ensure that a fluid is flowing through the system to prevent overheating.

Evacuated tube collectors vary widely in how they are constructed and how

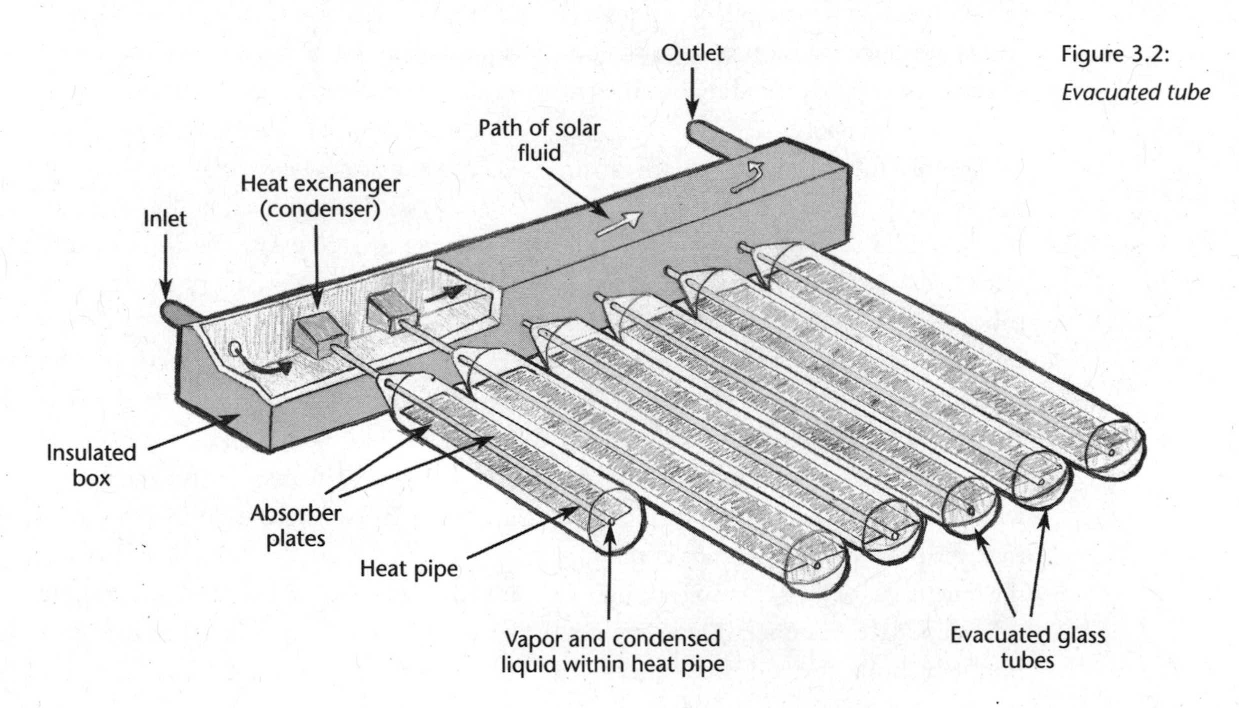

Figure 3.2: *Evacuated tube*

they heat a fluid. The principal distinctions are how many layers of glass they have and whether they heat the solar fluid directly or use a heat pipe.

Glazing

The collectors can be constructed of either a single or double tube of glass. In single-tube versions, the absorber plate and riser tube rest directly inside the vacuum. The configuration means that the solar radiation has to pass through only one layer of glass, but it requires a seal where the riser tube leaves the vacuum. This seal needs to be able to withstand the high temperatures experienced within the vacuum tube, but it must also remain pliable enough to withstand the expansion and contraction of the copper pipe extending out the end as it heats and cools.

Double-tube versions go by many names, Sydney tubes, Dewar tubes or Twin tubes, but all versions are principally the same. A tube is placed inside a larger tube, and the two are sealed together at the ends. The vacuum is then drawn into the space between the layers of glass. The absorber plate is either coated on the underside of the inside layer glass or placed within the vacuumed space. In the latter case, the heat will need to transfer through the inner layer of glass. Having the double tube eliminates the metal/glass connection, which may extend the longevity of the vacuum but requires energy transfer through two layers of glass.

Heat Transfer

Most evacuated tubes on the market today use a heat pipe to absorb and transfer the solar heat. A heat pipe is a sealed hollow tube that is filled with a small amount of fluid. This fluid can be water, an alcohol/water mix, ammonia or some sort of proprietary blend. A vacuum is then created inside the pipe and it is sealed. When the atmospheric pressure is reduced inside the pipe, the vaporization point of the liquid is lowered so that when the pipe gets hot the liquid vaporizes at a lower-than-normal temperature. When a liquid changes state to a gas, a lot of energy is absorbed. When a gas condenses back to a liquid, a lot of energy is released. This energy transfer when a material changes state is called latent heat, and because so much energy is transferred when the state is changed this makes heat pipes very efficient at moving heat.

The heat pipe is attached to the absorber plate and placed inside the vacuum tube and then extends out the top. The absorber plate construction and coating materials that are used in evacuated tube collectors are the same as those used in flat plate collectors. When sunlight strikes the plate, the pipe is heated and the liquid inside it evaporates. The hot vapor then rises to a heat exchanger in a manifold located along the top of the tubes. On the other side of this heat exchanger is the solar fluid, which absorbs the heat and circulates throughout the system. As the solar fluid cools the vapor, it condenses and drops back down into the pipe. Some models of evacuated tube collectors have an automatic overheat protection built into the heat pipe using a bi-metal switch. When this switch gets too hot, risking overheating the system, it changes shape, causing flow in the heat pipe to stop, interrupting the transfer of heat to the solar fluid.

Collector configurations that don't use a heat pipe are called direct-flow or flow-through evacuated tubes. Almost all direct-flow systems use a double-tube configuration, so there isn't a metal/glass connection involved. The riser tube containing the solar fluid passes down along the absorber plate, makes a U-turn and comes back out the same end it entered. Early evacuated tube versions passed the solar fluid directly through the tube, going in one end and out the other. This configuration was problematic because it required two seals, and they consistently lost their vacuum. Direct-flow configurations are typically more efficient than those using a heat pipe because the solar fluid is in direct contact with the absorber plate, and you eliminate any losses in the heat transfer process from the absorber plate to the heat pipe.

Other Characteristics

Some evacuated tubes use a getter to ensure that all of the gasses have been removed from inside the tubes. Commonly made of pure barium, the getter will absorb any residual gasses left over after the vacuum has been drawn. The barium will leave a shiny silver coating on the bottom of the tube. If the tube loses its vacuum, this barium coating will be exposed to oxygen and will become white or foggy, clearly indicating that the vacuum has failed and the tube needs to be replaced.

Since the tubes themselves are not structural components of the system, a rack is usually constructed to hold them in place. There is typically a bottom rail that all of the tubes can rest in and be secured. The top of the tubes enter into the manifold and are secured there. Most mounting kits come with bracing members to connect the bottom rail to the manifold. This assembly can either be flush mounted to the roof or raised using struts. Because the tubes are removable from the manifold, they can be installed after the rack has been constructed. Evacuated tubes that use a heat pipe must always be mounted at a pitch of at least 25 degrees to ensure that the evaporation/condensation cycle will flow.

ICS Collectors

ICS stands for Integral Collector Storage. In an ICS unit, the solar hot-water-storage tank is the solar absorber. The tank (or tanks) is mounted in an insulated box with glazing on one side and is painted black or coated with a selective surface. The sun shines through the glazing and hits the black tank, warming the water inside the tank. Some models feature a single large tank (30 to 50 gallons) while others feature a number of metal tubes plumbed in series (30- to 50-gallon total capacity). The single tanks are typically made of steel, and the tubes are typically made of copper. These collectors weight 275 to 450 pounds when full, so wherever they are mounted, the structure has to be strong enough to carry this significant weight.

Figure 3.3: *Tank type ICS collector*

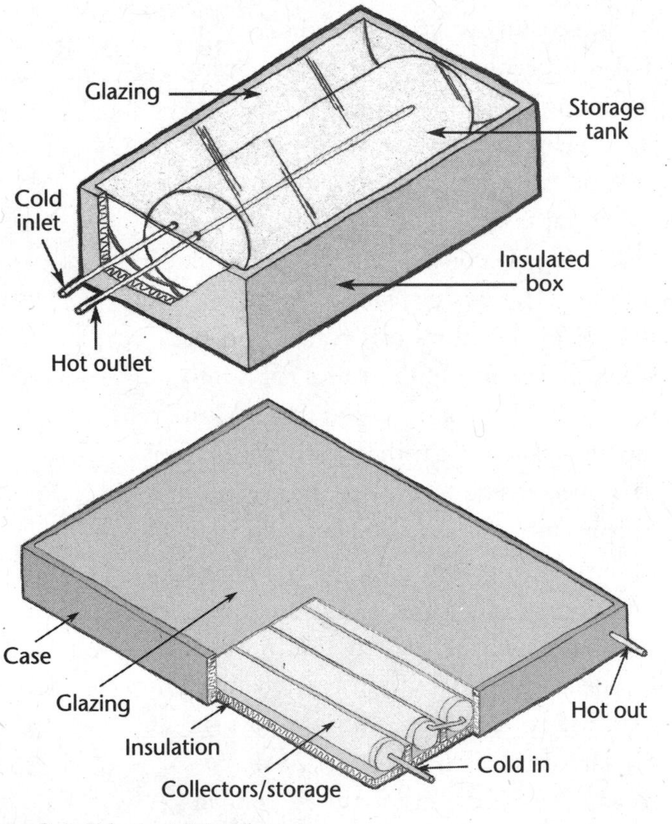

Figure 3.4: *Tube type ICS collector*

ICS collectors are widely used around the world in climates that never experience freezing conditions. They work great, given their climatic restrictions. They are a direct type of system, as the water you use is actually heated in the collector. They do not suffer from hard-water problems nearly as much as do flooded collectors (collectors with water in them at all times) because all waterways in the collector are very large and also because the collector never gets hot enough to precipitate minerals out of solution, and their simplicity makes them a very affordable choice.

ICS collectors are also used in seasonal applications such as campgrounds and summer homes, where they are used only during the warm months of the year and are drained before freezing conditions occur. Bob has sold many of these collectors with excellent results.

As mentioned above, these collectors are heavy, so if they will be mounted on a roof, make sure that roof is strong enough to hold their weight. You may have to reinforce an existing roof to make it acceptable. These collectors should always be tilted so they will properly drain.

The tube type of ICS collector will outperform the tank type because more surface area is exposed to the sun. Another advantage of the tube type is that its profile is much smaller, which affects aesthetics. On the other hand, tube type collectors cool off more quickly at night because of their larger surface area. On cool nights, the water stored in these collectors will cool off, so they lose efficiency. You can maximize this kind of system's efficiency by using as much hot water as you can during the day and early evening hours.

Another kind of collector system often classified as an ICS system is called a thermosiphon type system. These use a flat plate or evacuated tube collector mounted directly to a storage tank that is located directly above and attached to the collector. They use a heat transfer fluid that flows through the collector or uses a heat pipe to deliver heat from the collector to the storage tank. These systems are very popular in southern Europe, Asia and areas of Australia where freezing conditions never occur. These systems do not lose heat during the evening hours as quickly as the tank type collectors do because the storage tank is better insulated.

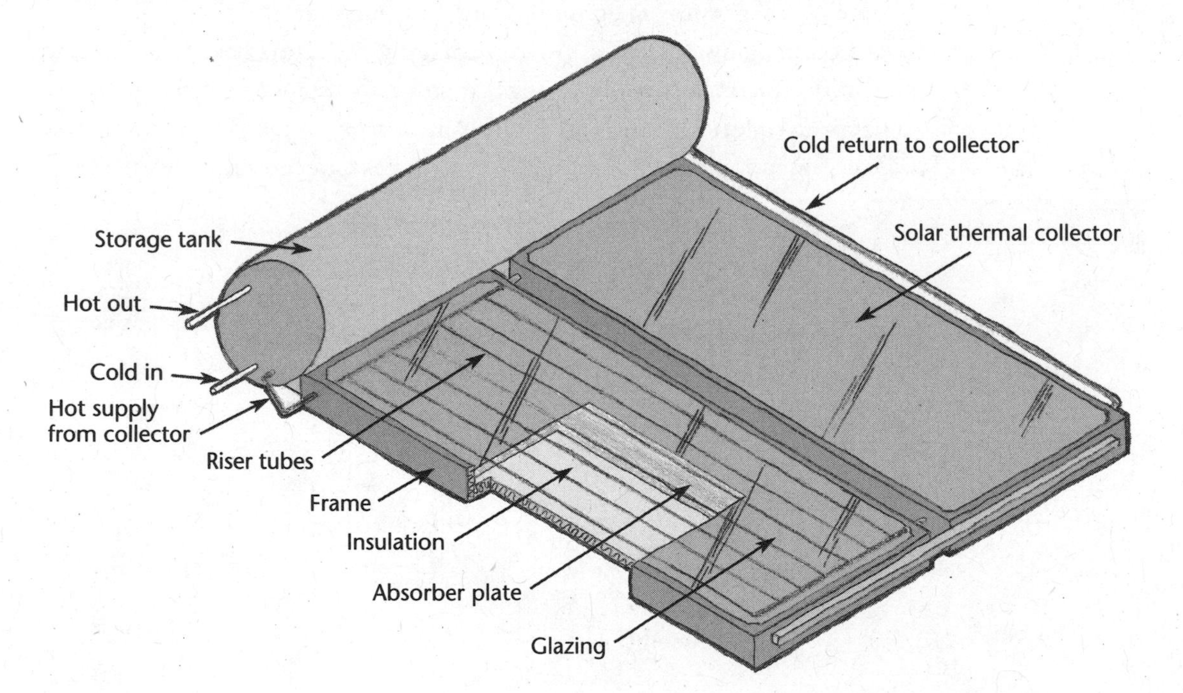

Figure 3.5: *Thermosiphon type ICS collector*

All ICS systems are plumbed in series, with the backup water heater acting as a pre-heater. Very often they can provide 100 percent of the daily domestic hot water when installed in hot, sunny climates or during the summer months when the sun shines most days.

Concentrating Collectors

Concentrating collectors use a reflective parabolic-shaped surface to reflect and concentrate the sun's energy to a focal point, where the absorber is located. Concentrating collectors use only direct sunlight; most other types of collectors can also use indirect sunlight. Concentrating collectors come in two types: imaging reflectors and non-imaging reflectors. A non-imaging reflector collector looks much like a flat plate collector, as the mirrors and receivers are mounted in a glazed frame of similar dimensions to flat plate collectors. Non-imaging reflectors do not track the sun but remain stationary while still retaining the ability to focus the sun's energy on the receiver as the sun moves across the sky. These collectors make up a very small percentage of the total collector market. To work effectively, imaging reflectors must track the sun. These collectors can achieve very high temperatures because the diffuse solar resource is concentrated on a small area. In fact, the hottest temperatures ever measured on the earth's surface have been at the focal point of a massive concentrating solar collector. This is sort of like starting a fire with a magnifying glass on a sunny day.

Imaging collectors have been used to make steam that spins an electric generator in a solar power station. This is commonly referred to as Concentrating Solar Power

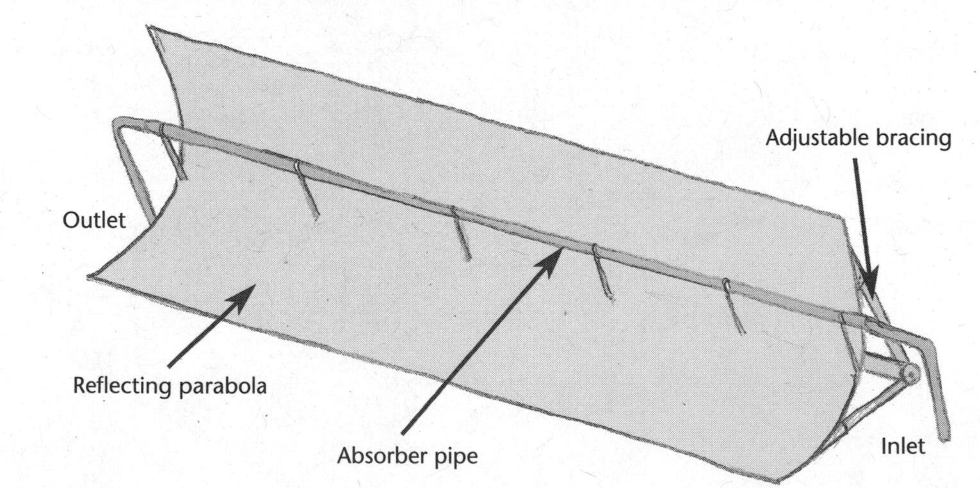

Figure 3.6: *Concentrating collector*

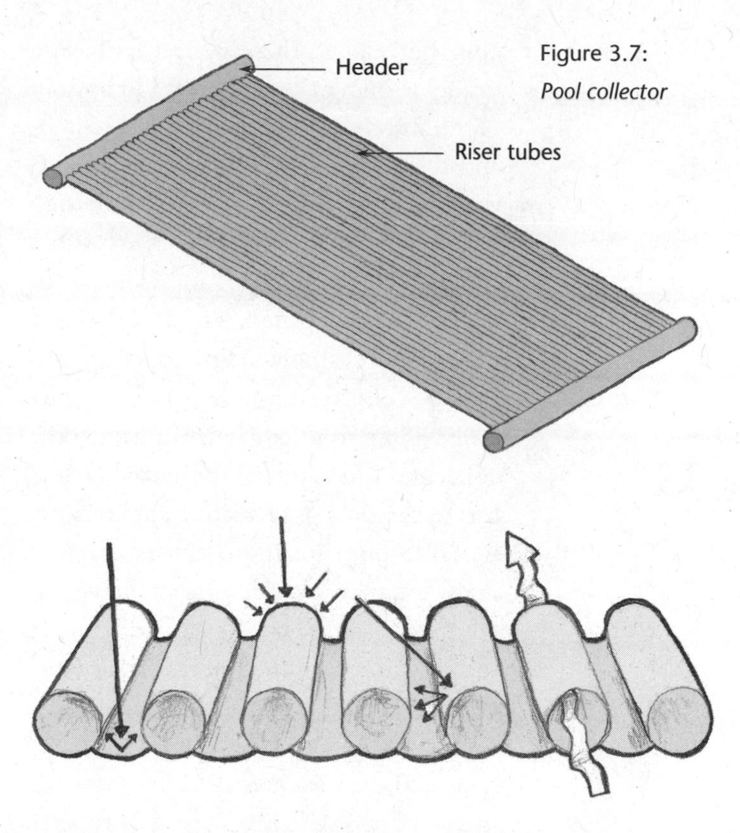

Figure 3.7: *Pool collector*

Figure 3.8: *Web-and-tube configuration*

(CSP). Large electricity generating plants using imaging concentrating collectors have been constructed in dry climates around the world with great success. A lot of research is being done in this field, and we can expect this technology to grow significantly in the near future.

Although there have been attempts to use imaging collectors in domestic water heating systems, no successful or durable products have been developed. Problems were encountered with the tracking mechanisms, the precision needed in the mechanisms and the durability of the reflectors and linkages. Recently a couple of firms have been releasing imaging collectors for small commercial and industrial use. Given the poor track record of imaging collectors, we are still unsure whether this second attempt will be a success. They are particularly vulnerable to snow and ice buildup in cold climates.

Pool Collectors

The single largest application of active solar heating systems in the world is in heating swimming pools. Special collectors have been developed for heating seasonal swimming pools: they are unglazed and made of a special copolymer plastic. Most are made of proprietary materials containing UV stabilizers to prolong their lifespan. The collectors don't have to be glazed because they are used only when it is warm outside, and they are heating water to a relatively low temperature that is just slightly above ambient. These collectors cannot withstand freezing conditions.

The absorber plate in copolymer collectors is an extruded mat that has many waterways, or risers, that are set close together. Many designs have been produced, but the design that has proven most successful is a web-and-tube configuration. With this design, the tubes are separated by a small web so they are not touching one another. Much like copper absorber plates, these collectors have a header running along the top and bottom of the absorber. The best designs, like web-and-tube absorbers, are made so that each riser is individually attached to the header, which allows the least back pressure within the collector and allows individual connections to be repaired, if needed. Because the risers are situated close together, this type of collector has a large wetted surface, adding to its efficiency. In fact, these collectors can outperform glazed collectors in seasonal pool heating applications.

Solar pool heating systems that use unglazed plastic collectors are direct systems in which the pool water is circulated through the collectors. This is the most efficient configuration as there are no heat exchanger losses. Most pool water contains additives such as chlorine, which is highly corrosive. This corrosive property of pool water makes copper absorbers or copper pipes noncompatible. This is one of the reasons the collectors are made of plastic, which is not affected by chlorine. The same holds true for the piping, which is typically PVC.

Plastic pool heating collectors are typically mounted flat on a roof. The collectors are held in place with a set of straps that go over the collectors but are not actually attached to them. The straps are often plastic-coated stainless steel and are threaded through special clips that are bolted to the roof. This method of holding down the collectors allows them to expand and contract on the roof without binding and makes them easy to install.

Comparing Solar Water Heating Collectors

People often ask us, "What is the best type of collector?" We would like to be able to give you an easy answer, but unfortunately there isn't one. We have found that many factors come into play when it is time for you to select a collector. It depends on the type of application, the climate of the installation, the amount of room available for the collector array, the budget for the project and other reasons too numerous to mention. What is more important than trying to find the "best" collector is being able to match the characteristics of the collector to the job that needs to be done and the limitations of the site.

Collector Efficiency

No matter which manufacturers you talk to, they will probably claim that they offer the most efficient collector. The truth of the matter is that they are all wrong. There is no such thing as the "most efficient" collector. For instance, consider evacuated tube collectors. Since their conception, much has been said about their performance as they are commonly heralded around the solar industry as a more efficient collector. If you were to look just at the collector design, that's an easy assumption to make. Normally, if you reduce heat loss, in this case by the vacuum of the tubes, you would increase efficiency. So it may sound counterintuitive if we tell you that pool collectors, which are simply plastic tubes placed out in the sun, are actually more efficient for some applications even though they are not as well insulated. It takes more than a clever design to produce hot water. It takes a system that is appropriately suited to what you want it to do.

The measure of a collector's efficiency should really be how it performs when put to use. The best way to measure that is through an independent testing organization. One of the good things that came out of the late '70s solar boom was the Solar Rating and Certification Corporation (SRCC). The SRCC rates and certifies many of the collectors on the market today. It is the most common and reliable source in the United States for independent information about solar collectors. The SRCC does not perform the required tests on the collectors. The test was developed by American Society of Heating, Refrigeration and Air-conditioning Engineers, and is performed at accredited testing facilities. The SRCC uses the test results when rating the collectors We strongly suggest buying collectors that they have certified. Not only does the test calculate collector performance and efficiency, it also tests for durability and reliability. Both are critical for determining the value of a collector. The results are free to the public and can easily be accessed online at solarrating.org. Using the SRCC gives us good, solid standardized data for comparing collector performance.

When rating a collector, the test measures the amount of heat, in Btu, that it will produce, based on a certain amount of radiation that shines on the collector. The testing facility usually does this with big lights to ensure consistency between tests, but some facilities conduct the tests outdoors using real sunlight. Because the solar resource is inconsistent, three conditions are considered: clear day (2,000 Btu/ft^2/day), mildly cloudy (1,500 Btu/ft^2/day), cloudy day (1,000 Btu/ft^2/day). The conditions mimic how the amount of sun will vary depending on location and climate. As a second variable, the test will alter the temperature at the site. This is actually the difference of the temperature of the fluid going into the collector (inlet temperature) and the temperature outside (ambient temperature). This gives you a measure of how hot the fluid is that you are trying to heat and how cold it is outside. Figure 3.9 graphs the ability of each type of collector to convert sunlight into usable Btu for all of the temperature

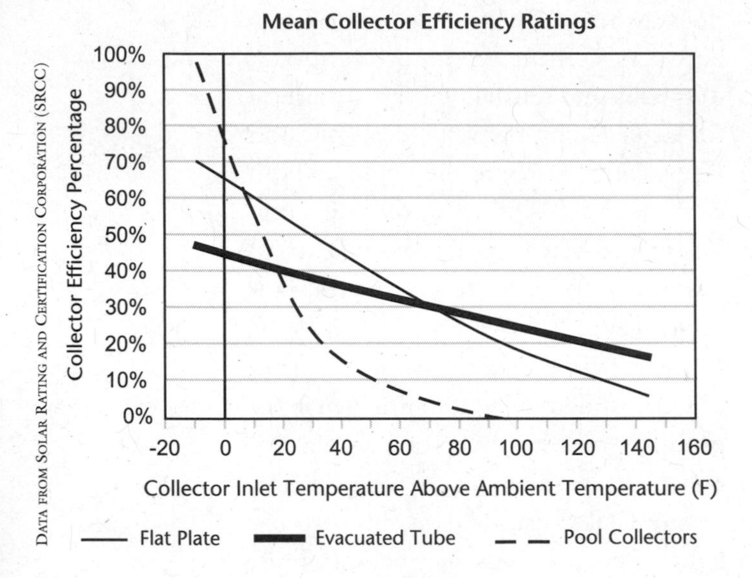

Figure 3.9: *Mean collector efficiency ratings*

variables. The data is an average of all three sun conditions and was taken from a sample of ten manufacturers of each type of collector to provide a measure of overall performance.

As you can see, when there is very little difference between the inlet temperature and the ambient temperature, the pool collectors are significantly more efficient than both flat plate and evacuated tube collectors. Does this make them the most efficient collector? No. It simply means that they are better during some conditions. Similarly, the flat plate collectors are more efficient when the inlet/ambient temperature difference is between 10°F and about 70°F. After that point, the evacuated tube collectors become more efficient. The efficiency of the collector is entirely contingent on where and how it is being used.

The question now should be, where does my situation fit into this? For most domestic water and space heating applications we are trying to get our fluid up to 120°F–140°F. Let's consider an example in which you have a system design to heat your domestic hot water. Let's say it is 50°F outside and the fluid returning to your collector is 100°F. In this condition, you would look to the point at 50°F on the graph. Flat plate collectors are about 40 percent efficient, and evacuated tube collectors are around 34 percent. That's quite a difference in performance, even on a relatively cold day. If you are properly dumping the heat, the inlet temperature on most residential applications is usually 100°–110°F at most. However, at that point you will not need much more to reach your desired temperature.

We have found that for most residential water and space heating conditions, flat plate collectors will outperform evacuated tubes. Now, if you needed really high temperatures, say higher than 160°F, then evacuated tubes might be the right collector for the job. Like we said before, it all depends on where and how it is being used. Pick the right tool for the job.

The second claim made for evacuated tube collectors is that they are better collectors during cloudy conditions. Figure

3.10 graphs the efficiency ratings for all three SRCC conditions, including cloudy, low-sun weather. As you can see, the point where the collectors' efficiency ratings cross is less than the average, signaling an increased efficiency. However, they are still not more efficient than flat plate collectors in most temperatures.

Additionally, you need to consider the value of a system being better at harvesting a decreased resource. If there isn't much solar radiation to gather in the first place, being slightly better doesn't amount to a whole lot of Btu. More of a little bit is still only a little bit.

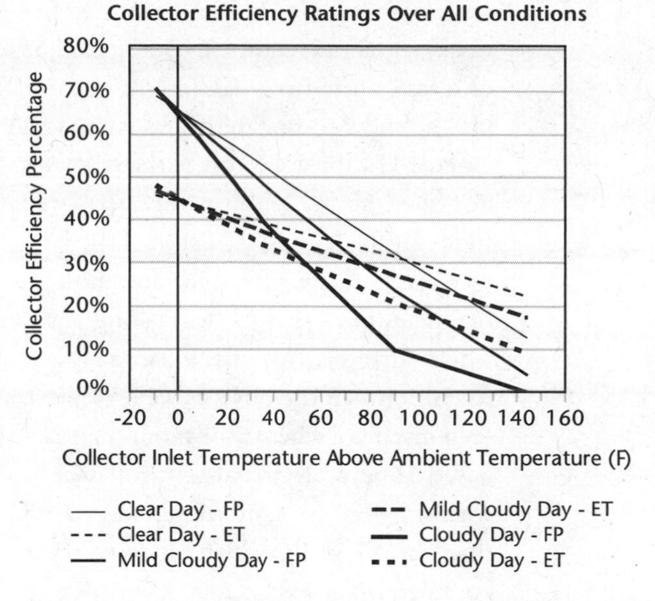

Data from Solar Rating and Certification Corporation (SRCC)

Figure 3.10: *Collector efficiency ratings over all conditions*

Durability

Although much of the industry debate has centered around efficiency, as it is the obvious selling point, we have tended to focus more on the long-term performance of collectors than on the short term. The system you install today should be designed to last at least 40 years. There is no substitute for quality. The collectors on Bob's home were installed in the early 1980s and are still in great condition.

An easy indicator of the quality of the collector is the warranty. Most come with a 10-year warranty, although some have stretched this to 15 years. You might also want to check into the manufacturer. How long has the company been in business? A few companies have been making collectors since the '70s and '80s and have continued to put out long-lasting products. You may also want to find out where the manufacturer is from. We have several companies here in the Midwest that we like to deal with because they are local. This not only cuts down on shipping and transportation costs; it also allows us some flexibility if we need something in a hurry.

Climate and Location

In some climates, such as ours, snow is a significant factor in collector performance. Snow can accumulate on the collectors and diminish the solar resource. Flat plate collectors have a distinct advantage over other types in that they shed snow very well when installed in climates that experience significant snowfall. The large pane of glass on the front will lose heat, causing the snow to slough off. An issue with evacuated tube collectors is that they do not shed snow. Because the evacuated tubes are such good insulators, little heat escapes, and the snow that accumulates on the tubes can stick for a long time. Their surface is also irregular, so snow packs between the tubes as well. We have seen instances where roof-mounted evacuated tube collector arrays got packed with snow in the early winter and stayed that way till spring, which rendered them completely useless for a good portion of the year. The lesson here is to always mount evacuated tube collectors at a significant angle if used in a climate that experiences snow. Also, never flush mount evacuated tube collectors in a climate that experiences snow.

Absorber coatings will influence the performance of a solar thermal collector, and the amount of this effect may be more critical depending on the climate the collector will be placed in. Some types of absorber coatings have a higher performance rating than others. In climates that experience a high number of sunny days per year and also experience relatively consistent warm temperatures, collectors can perform well with less efficient absorber coatings, whereas in climates with fewer sunny days per year and colder temperatures, collector coatings with higher efficiencies work best.

You may also need to consider the density of the absorber area compared to the overall footprint of the collector. Manufacturers and certification agencies use the terms *gross collector area, net aperture area* and *absorber area. Gross collector* is the entire area including the frame; *aperture area* is the size of the glass; and *absorber area* is the amount of surface that will actually absorb solar radiation. In flat plate collectors virtually the whole collector area equals the absorber area, with only about an inch around the edge being frame and not absorber. In an evacuated tube collector, there is space between each tube that is not absorber area. In most instances a flat plate collector will take more than 25 percent less space than an evacuated tube collector that has an equivalent absorber area. This may be an important factor because a common limiting factor in siting collectors is the amount of available mounting space. In other words, you want to have enough space for the amount of collector area you want, so to get as much heat as you can, you want to have a type of collector that has the best gross-to-net absorber ratio.

Note that the amount of solar energy that falls on a square foot of the earth is a constant and cannot be changed by the type of collector used. The primary way to increase the amount of energy collected is to increase the absorber area.

By now it has probably become clear that we are not completely impartial when it comes to collector selection. It is difficult to remain unbiased while still trying to provide the knowledge we have gathered through years of experience installing, maintaining and designing these systems. We have seen the best performance from flat plate collectors, and we want you to have the same success. We have no doubt that evacuated tube collectors have a secure place in the solar thermal industry, especially in high-temperature applications. For instance, the emerging solar thermal-powered air conditioning systems that use evacuated tube collectors to drive single- or double-effect chillers hold great promise. However, for the majority of domestic water and space heating applications, flat plate collectors have a proven track record.

Air Collectors

Up to this point, all the kinds of collectors we have talked about have used a liquid as the heat transfer medium. Air can also be used as the heat transfer mechanism in a solar collector. Air collectors are flat plate collectors and share all the same characteristics of liquid-type flat plate collectors in size and construction. Instead of an absorber plate made of copper piping and copper fins, the absorber plate in an air collector is typically made of a solid sheet

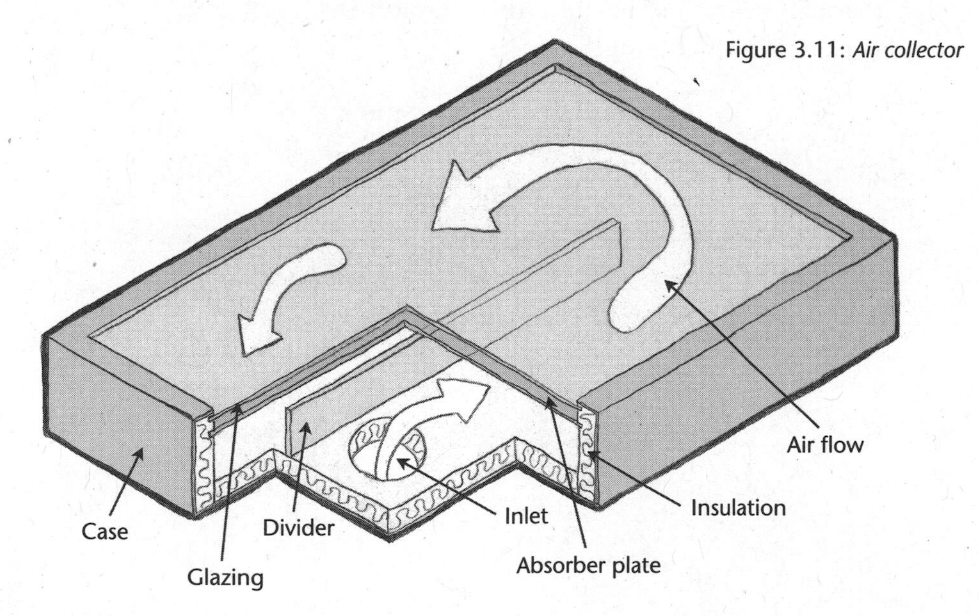

Figure 3.11: *Air collector*

of aluminum. The aluminum absorber plate is coated with a selective surface or black paint and is usually dimpled to increase efficiency. When the sun shines on the absorber plate, it gets hot. Air is drawn from the building and is blown across the back of the absorber plate and heated. The hot air is then delivered to the building through ductwork. A blower circulates the air through the system.

Air collectors can come in standard sizes that are very similar to flat plate collectors and they can also be site built to fit a particular building. We personally manufactured site-built air collector solar space heating systems in the early and mid-1980s and had great success. These collectors should be less expensive than liquid-type solar thermal collectors because the absorber plates in them are much simpler and are typically made of aluminum, which is less expensive than copper. These collectors can produce the same Btus per square foot as any other collector, so they can have very attractive returns on investment.

Another kind of air collector is a transpired collector. This collector is an unglazed flat plate collector. It draws fresh air through tiny holes that are punched in the absorber plate. This fresh air is heated as it passes through the collector and is delivered into the building. This kind of collector is used for heating makeup air in buildings that require many air changes per day. As these collectors are even simpler, with no glazing and very simple frames, their cost is very attractive, and these systems have very good returns on investment.

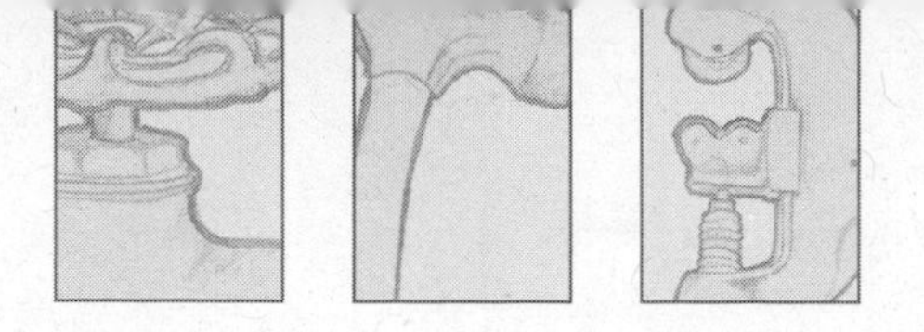

4

OTHER SYSTEM COMPONENTS

THIS CHAPTER WILL DETAIL a number of the other components that are used in solar water heating systems. This is not an exhaustive list of components; it includes only those that are most commonly used. Components that are designed for solar systems are typically preferable.

Storage Tanks

Because of the path of the sun, there is a limited time in which we can harness solar energy. But we aren't willing to use hot water only when it is sunny. For instance, we mostly shower in the morning or at night. Therefore, all solar water heating systems will require some form of storage tank, where the solar-heated water is stored until needed. These tanks typically range in capacity from 40 gallons to 120 gallons but can be much larger with space heating systems. Determining the right size of storage tank will be covered in Chapter 7. As with most things, there are a number of options to consider when choosing a storage tank. The type of system you are planning will determine some of the specifications of your storage tank.

But first, a quick discussion on tanks in general. The most common and traditional tank used today is a steel tank. The steel tank is encased in foam insulation to reduce heat loss and has a light-gauge steel jacket on the outside to protect the insulation. It looks just like a regular water heater. The inside of the tank should be coated with an enamel layer, often called glass lining, which is typically baked on. This lining helps reduce corrosion and significantly prolongs the life of the tank. High-quality steel tanks should also be fitted with an anode, or sacrificial, rod, which is screwed into a fitting on the top of the tank and extends down into the tank. An anode rod helps reduce tank corrosion by rusting before any of the system components do. Anode rods actually wear away, and their life expectancy varies depending on the conditions at your location. The anode rod should be checked every five to ten years. Steel tanks typically last 15 to 30 years, depending on the environment at your location and the quality of your water.

Fiberglass and plastic tanks are the new kids on the block. Modern developments in fiberglass and thermoplastic technology have enabled engineers to create a cost-competitive alternative to steel tanks. These tanks are constructed much like the steel tanks except that the tank itself is made of fiberglass or plastic and the jacket is plastic. They have a huge advantage over steel tanks because they will not deteriorate because of rust or corrosion. When using these types of tanks, a few special precautions should be taken. First, when screwing fittings into the tank, it is important not to over-tighten the fittings. Many installers tend to crank fittings very tight when working on steel tanks, and there is no problem with that, but on plastic or fiberglass tanks you can break the fitting by applying too much force. So be sure to use plenty of pipe dope, sealant or Teflon tape on the fitting, and do not over-tighten. Second, install a vacuum breaker on the top of the tank to facilitate safe drainage. Whenever a tank is being drained and the tank is sealed at the top (all faucets above it are closed), a vacuum is created at the top of the tank. This is no problem with a steel tank, as steel is strong and rigid. Although a fiberglass or plastic tank is very strong in relation to outward pressure, it is weak when it comes to the inward pressure that would be caused by having a vacuum inside the tank.

If you choose to install a plastic tank, be sure that the plastic is rated to handle prolonged exposure to high temperatures. Most types of plastics will deteriorate under the constant hot conditions.

Tanks with Built-In Heat Exchangers

Some tanks are constructed with a heat exchanger as an integral part. Commonly referred to as indirect tanks, these have been popularized by the heating and plumbing industry as a means to heat domestic hot water with a separate hydronic heat source, typically a boiler. It used to be the case that tanks with internal heat exchangers were a specialized product to the solar thermal industry, making them very expensive. However, their application in the more traditional heating and plumbing field has expanded their usage and, consequently, driven the price down significantly.

There are several indirect tank configurations for you to consider. The most common setup is to have a knurled or rifled copper or stainless steel coil of tubing placed inside a steel tank. This provides the greatest surface area in contact between the heat exchange coil and the domestic water, increasing efficiency. However, over time the textured surface of the coil may become coated with scale and mineral deposits, decreasing efficiency. This is especially true in hard-water situations. As long as the tank is flushed at regular intervals or installed in systems with soft or softened domestic water, it will be successful.

To avoid issues with scaling, some manufacturers of indirect tanks place the coiled heat exchanger on the outside of the tank. An illustration of the wraparound method can be seen in Figure 4.1. This technique reduces the amount of surface area that is in contact with the water you are heating but ensures consistent and prolonged efficiency of the tank. Instead of a coil, a few manufacturers use a fluid-filled jacket surrounding the tank. Essentially, the domestic water tank is submersed inside a slightly larger tank. In areas with hard water either one of these methods would be preferable.

Some tanks have an internal heat exchanger that is removable and is accessible through a bulkhead fitting on the side of the tank. You can also buy a heat exchange "wand" that can be installed through the top port of standard electric or gas water heater. A removable heat exchanger allows for easy access and cleaning, but may limit the total amount of heat exchange surface area that can fit into a single tank.

Indirect types of solar storage tanks have several advantages. First, installation is significantly simplified because you don't

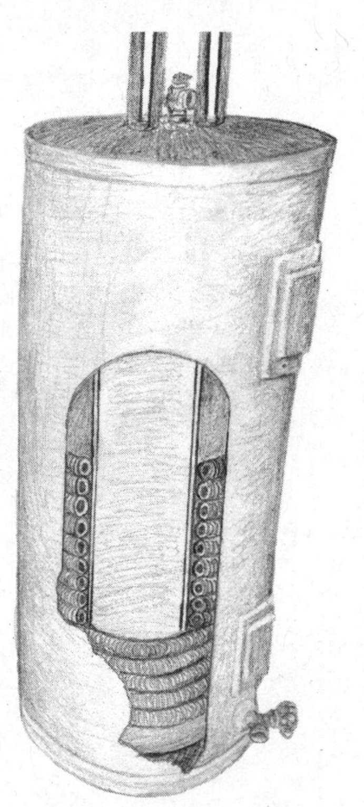

Figure 4.1: *Solar storage tank with wraparound heat exchanger*

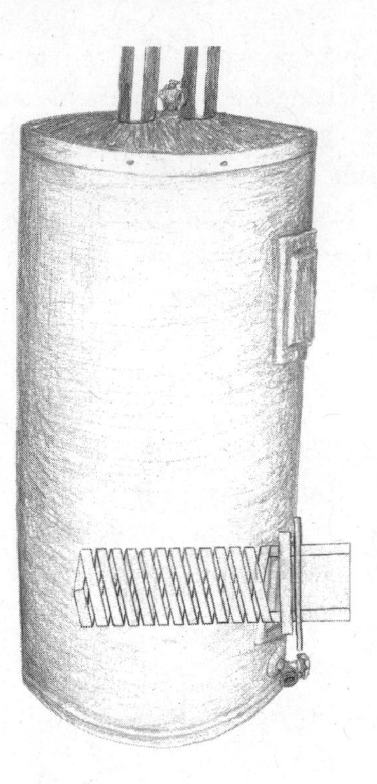

Figure 4.2: *Solar storage tank with internal heat exchanger*

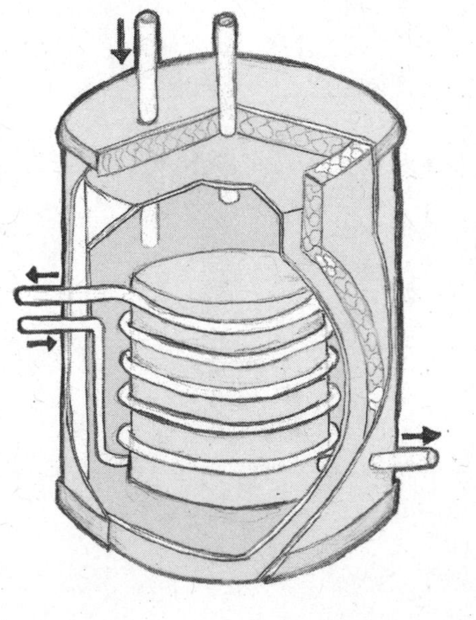

Figure 4.3: *Drainback tank with internal heat exchanger*

have to do the extra work of mounting and plumbing an external heat exchanger. Though indirect tanks are more expensive, you must remember that you are eliminating the cost of an external heat exchanger and the cost to install it. Second, the system will require only a single main circulating pump. Most external heat exchangers require a pump on the solar side and on the domestic hot water side. The second pump results in not only an initial installation cost but also increased parasite loads and utility bills over time. Finally, the tanks can also save space where quarters are tight. After years of working in cramped utility rooms, and sometimes closets, we have developed an understanding of the value of floor space. Sometimes, this type of tank will be the only option.

Tanks without Built-In Heat Exchangers

Storage tanks without built-in heat exchangers are the most common kind found in solar water heating systems. These can be made of steel or fiberglass. To function as a solar hot-water-storage tank, the tank has to be insulated and have suitable fittings to allow proper connections of the water supply, the heat exchanger and the drain. A number of manufacturers build storage tanks specially designed for this type of application, similar to water heating tanks but with extra ports to make it easy to install the solar plumbing. It is also possible to use a traditional water heater tank. These tanks are readily available at competitive prices and come in both fiberglass and steel. If you go this route, it is best to use an electric water heater. A gas water heater tank will most likely have a flue down the center, which can lead to more standby heat loss, as well as providing more surface area to develop leaks. More information on how configure a traditional electric water heater into a storage tank will be covered in Chapter 8.

Sometimes folks want to buy a used tank when installing a system. If you choose to do that, you should inspect the tank carefully for corrosion. In most cases the used tank was taken out of service either because it leaked or was old and ready to leak. We do not suggest using used storage tanks.

Drainback Tanks

Drainback systems need specialized reservoir tanks in addition to the storage tanks. They are like small solar storage tanks and usually hold about 5 to 20 gallons. Some drainback tanks have a heat exchanger built into them. Many have a sight glass built in to the side of the tank for monitoring the liquid level inside the tank. These specialized tanks are available from collector manufacturers and solar specialty shops. As with any other tank, you want to look for glass-lined steel or fiberglass tanks with adequate insulation.

Solar Space Heating Storage Tanks

Solar space heating systems are just overgrown solar water heaters. Being overgrown, they need an overgrown storage tank — think big. The size will vary depending on house size and climate. It is not uncommon for these storage tanks to be in the 300-gallon to 500-gallon range. To achieve the required storage size you can use either one large tank or several smaller tanks plumbed together. One large tank will always be preferable. These tanks are made of steel, stainless steel, fiberglass or, occasionally, concrete. There have been many attempts over the years to build storage tanks in place and line them with plastic or rubber. These do not work and will fail. The liners in these tanks are not designed to withstand temperatures commonly encountered in the system. They will often become brittle and eventually leak and crack.

One big problem with large storage tanks is getting them into a building and through doorways. Some manufacturers have made tanks that come in pieces and are assembled in place. These are usually made of fiberglass and either bolt together or are cold welded with epoxy. We suggest selecting a cold-welded tank that is made of heat-resistant fiberglass capable of withstanding at least 180°F in continuous operation.

Heat Exchangers

Liquid-to-liquid heat exchangers are used in a solar water heating system to transfer the heat from the solar fluid into the domestic water.

Most heat exchangers are single walled. A single-walled heat exchanger has a single membrane between the two fluids. This membrane could be made of copper, stainless steel, or (in very specific circumstances) Pex. Double-walled exchangers afford an extra layer of protection to the potable water in case the heat exchanger ever develops a leak. Many have a gap between the layers, commonly referred to as "positive leak detection," so that a leak in a wall will clearly be visible, and the heat exchanger can be repaired or replaced.

Figure 4.4: *Plate heat exchanger*

Single-walled heat exchangers are typically more efficient because the heat needs to be conducted through less material. However, some municipalities may require a double-walled heat exchanger whenever domestic water is involved in the exchange. Rules requiring double-walled heat exchangers are often a carryover from the early years of our industry, when toxic liquids were used as heat transfer fluids. There is no need for a double-walled exchanger if a nontoxic solar fluid is used.

Heat exchangers can be constructed using either pipes or plates. Two fluids are passed next to each other, separated by a membrane. Heat transfers across the membrane from one fluid to the other. Always plumb heat exchangers in a counter-flow arrangement, as this is the most efficient way to transfer heat. This means that the two fluids will need to flow in opposite directions when passing through the heat exchanger. This allows for the greatest difference in temperature between the two fluids, which increases the rate of heat transfer.

A plate heat exchanger consists of a number of plates spaced apart and capped around the sides. Separate waterways are designed into them to allow the different fluids to pass through adjoining spaces in a zig-zagging counter-flow pattern. This back-and-forth arrangement means that plate heat exchangers always require a pump on both waterways, as they will not convect. This type of exchanger can pack a large amount of surface area into a small package. Because heat transfer is directly related to the amount of surface areas that are available, you are unlikely to have too much heat transfer surface area. Because they have so much surface area in such a small package, plate heat exchangers also have very small waterways. Consequently, in hard-water conditions, scale can easily build up, causing reduced heat transfer or clogging. When installing one of these, you may want to place boiler drains and isolation valves on the domestic water side pipes to provide the means to periodically flush out the heat exchanger. Remember to always use a nontoxic solution when doing so. We prefer a product that is enzyme-based, making it biodegradable, non-corrosive and safe for septic systems.

A tube-in-shell heat exchanger is essentially a smaller pipe (or pipes) inside a larger pipe or tube. One fluid is circulated through the inner pipe(s) and the other fluid circulates through the outer pipe. The material of the inner tube separates the fluids. Tube-in-shell heat exchangers come in either single-wall or double-wall versions. These heat exchangers can be configured in straight lengths or coiled. The coiled configuration always requires a pump like the plate exchangers, but the straight ones can thermosiphon on the water side if properly designed for that purpose.

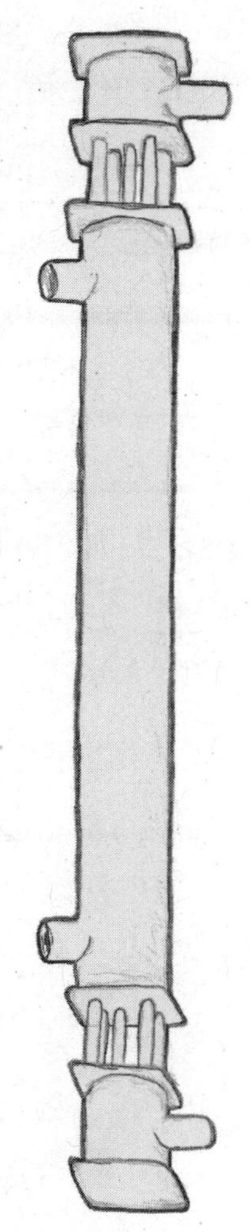

Figure 4.5: *Tube-in-shell heat exchanger*

Pumps

Pump Construction

It is important that all the components within a solar loop are made of compatible materials. When different materials are mixed in a plumbing circuit, galvanic reactions can take place that speed up oxidation and deterioration of the components. We prefer to use brass, stainless steel or bronze pumps in the solar water heating system, both in the solar loop and for all potable water plumbing. Some installers choose to use a cast-iron pump on the solar loop because it is often significantly less expensive. Though this may lead to corrosion, if the pump is separated from the piping loop through pump flanges, the gaskets often provide enough separation to prolong their use. The impellers within the pump are almost always made of stainless steel or plastic, so they are not a problem. It is the pump housing that is critical.

AC Pumps

The traditional type of pump used in a solar water heating system is a 120-volt AC pump. These pumps are readily available in a variety of sizes and are also used in traditional hydronic heating systems. Because these pumps are mass produced, they are less expensive than specialty pumps designed for the solar industry. There are many manufacturers of these pumps, which come in a variety of configurations. These pumps are reliable and last for a long time. Because of their low price, many of these pumps are not field repairable, so when they break, they are just replaced. A few manufacturers make pumps called cartridge circulators. These have a cartridge inside the pump that can be replaced without having to take the whole pump housing (called the volute) out of the system when doing a repair. Some AC pumps have multiple speed settings so that the flow rate can be optimized for the system.

Figure 4.6: *AC pump*

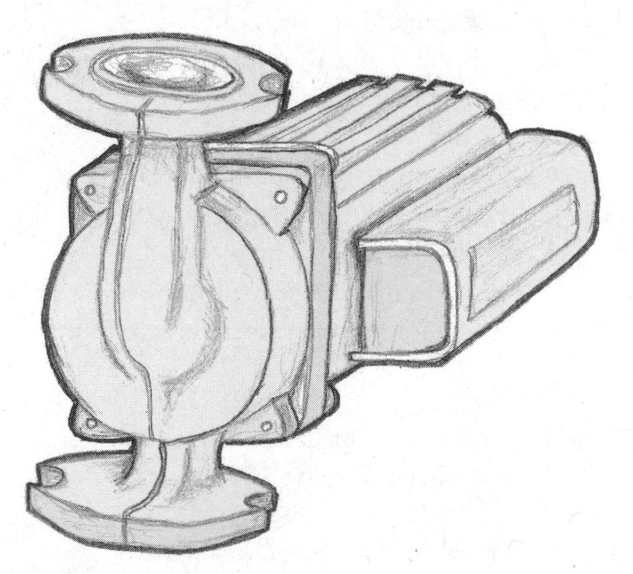

DC Pumps

Many solar water heating systems are installed with a DC pump so that it can be powered directly by a solar electric or photovoltaic (PV) module. DC pumps come in three configurations: brush, brushless and electronically driven. This designation relates to the type of motor that drives the pump.

Brush-type motors use brushes made of carbon that contact the commutator, which is a cylinder in the motor. These brushes wear away over time and need to be replaced periodically. Brushless motors are electronically commutated and, obviously, have no brushes. Brushless motors work better in photovoltaic direct applications, but brush-type motors can work fine. Brush-type motors have a harder time starting up when powered directly from a PV panel. They will start, but not as quickly as brushless ones.

A linear current booster (LCB) can be installed between the PV collector and the brush-type motor to help the brush-type motor start more easily. LCBs need to be properly sized to match the PV collector and the pump, so do your homework if using an LCB.

Electronically driven DC pumps fall into the category of brushless pump, but they don't have a conventional motor. These pumps use electronics to spin the impeller. Several brands of this type of pump have been developed specifically for the solar water heating industry, where low to moderate flows are required. One model even has built-in circuitry with maximum power point tracking (MPPT), which adjusts the voltage and current coming from the PV module to maximize the amount of power it is producing. Neat stuff.

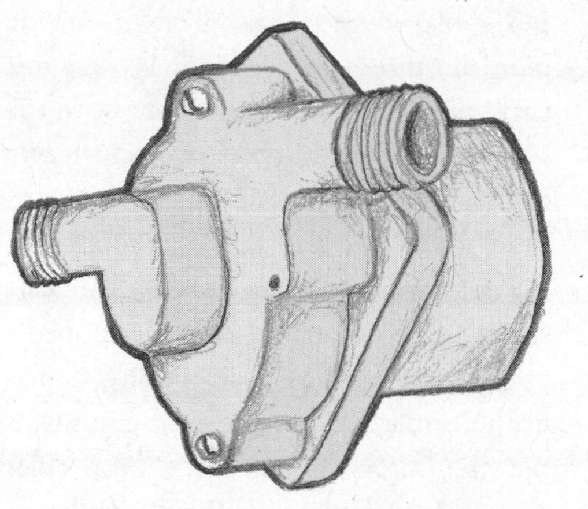

Figure 4.7: *DC Pump*

Piping and Pipe Insulation

Piping

The piping in a solar loop is subjected to a wide range of temperatures, varying from 300°F to –30°F (or lower). Copper tubing is the best kind of pipe to use for the solar loop. It can withstand this temperature range and is very durable and easy to install. Copper tubing comes in various grades and is classified by wall thickness and rigidity. Soft copper (annealed) can be bent; hard copper (drawn) is very rigid. The heavier the wall thickness, the more rigid it is. Type M copper tube is thin walled; type L is medium walled; and type K is heavy-walled. We suggest using type L for the solar-loop piping. Use type K for underground piping runs. We suggest using hard copper in all instances except underground piping runs. All types of copper pipe have the same exterior dimensions, so all fittings are made the same and come in one standard size for use with all types. However, the exterior diameter is actually 1/8" larger than its nominal designation.

It is important to use only copper pipe, or in some cases stainless steel, for the hot supply pipe in the solar loop. People often want to use a flexible product such as Pex tubing or rubber hose. These products will not last very long and will deteriorate well before the system wears out. In his repair business, Bob had to replace a lot of hoses that were used in the 1980s. Not a single installation from those days that used rubber hose is still operating in Wisconsin (to our knowledge). Pex tubing will also fail in a very short time. We know of several attempts to use Pex tubing for the hot supply line, and the failures were always within the first year. It is possible to use Pex tubing for the return line back to the collectors, but it is important to terminate the Pex at least 10 feet from the collectors. The

Pex must never be used for an outside pipe run unless it is buried. We suggest that you consider Pex only for use in the solar loop when burying the return line out to a ground-mounted array.

There is one alternative to copper for use in the solar loop. In the past few years several solar companies have developed a corrugated stainless steel piping that is both flexible and durable enough to handle the high temperatures of a solar system. Its flexibility simplifies installation and significantly reduces labor costs. It typically comes in a long roll with both the supply and return lines encased in high temperature insulation. Often a sensor wire is also sandwiched in the center of the insulation.

Though the product has now been around long enough to have a proven track record, it has a couple of drawbacks. The corrugation that makes it flexible creates increased surface area inside the tubing and results in significantly more resistance to flow, or friction head. In some cases this may require the use of a larger pump and cause increased electricity operating costs. Be sure to factor in the friction losses when selecting and sizing the pump. At this point it is also more expensive, although much of this is offset with savings in labor. Overall, it is still a good product and a welcome addition to the solar professional's toolkit.

Because of its flexibility, corrugated stainless tubing should never be used on a drainback system. The dips between the stainless links and sagging between pipe hangers will create pockets where solar fluid can accumulate. In a pressurized system, it may be necessary to flush or blow out the solar loop when replacing the fluid to ensure that all of the old liquid has been removed from the pipes.

Pipe Insulation

Pipe insulation comes in a wide variety of materials and specifications. Most kinds will not withstand the temperatures experienced in a solar loop; most plastic or rubber pipe insulation will melt right off. Only a few kinds of pipe insulation will work, and these are made specifically for high-temperature situations. We recommend that insulation should have a minimum continuous temperature rating of at least 250°F on the hot pipe coming off the collector to the heat exchanger, and a rating of at least 180°F on any other pipes on the solar loop. The material it is made of is not critical as long as the temperature rating meets the minimum standards. This will ensure that it retains its insulating value.

Fiberglass pipe insulation is great for all interior pipe runs but is unacceptable for exterior runs or buried runs because it will soak up moisture, become saturated and lose its insulation value. Some lower-temperature rubber and EPDM types will work, but unless intended for high temperature and exterior applications, they will eventually become hard and brittle and can wear away, especially when exposed to the elements.

It is always best to place a jacketing material on exterior insulated pipes. It is a tough world out there, and exterior insulation is subjected to UV radiation, rain, freezing conditions, insects and birds. Installing a jacket on all exterior pipe runs will greatly prolong the life of the insulation. Aluminum jacketing has proven to last the longest, but there are some new vinyl types that appear to be promising. PVC jacketing is also commonly used, but we have seen several issues with longevity in exterior applications, especially on the joints and fittings.

Solar Fluids

The best solar fluid is plain water. All other fluids are less efficient at transferring and holding heat. Because water is the best heat transfer fluid, many solar water heating system designs use it as the heat transfer fluid. The good and bad traits of the different types of systems will be discussed in Chapter 5.

Pressurized antifreeze systems use some type of fluid that will withstand freezing conditions. Many different fluids have been tried. Some examples are propylene glycol/water mix, ethylene glycol/water mix, synthetic oil and silicone oil. Of all the examples, only the propylene glycol/water mix is acceptable. The ethylene glycol/water mix is toxic; it will also deteriorate very quickly and damage your system. The synthetic and silicone oils require specialized components within the solar loop because the oils will dissolve all rubber and plastic components and seals, and they are just poor conductors of heat.

Propylene glycol/water mixes are the industry standard today for heat transfer fluids in solar water heaters. Propylene glycol is essentially nontoxic and biodegradable and is a relatively stable product, even at high temperatures. Be aware, though, that not all propylene glycols are created equal. These glycols come in various formulas and you need to carefully choose the appropriate formulation. You want to choose a product that is formulated to withstand at least 350° F of continuous usage without breaking down. Most major manufacturers make a propylene glycol product that is rated up to 370° F. Make sure you select the correct blend. In addition, many glycol products also have additives mixed with the glycol to help stabilize the product, inhibit corrosion and make it more compatible with system components.

You never use 100 percent glycol as your solar fluid; always mix it with a certain proportion of water. You want to have the highest percentage of water in the mix as possible and still have the protection you need for your particular climate. Dilution charts provided with the product show the necessary concentration of glycol in the mix to protect the system at various temperatures. Figure 4.8 is an example of a dilution chart.

These dilution charts show two types of protection: freeze and burst. As glycol/water mixes cool, the fluid becomes thicker and harder to pump. At a certain point ice crystals start to form in the solution, and the fluid will have a jellylike consistency but will continue to protect the piping from bursting. The freeze temperature on the chart shows the temperature at which the ice crystals start to form. The burst temperature on the chart is the temperature at which the fluid will begin to freeze solid and will burst a rigid pipe. For instance, a high-quality glycol at a 50:50 mix of water and glycol will provide freeze protection down to –34°F and burst protection down to –70°F. When choosing a dilution for your climate, you should choose one that is at least 10°F below the lowest temperature ever recorded at your location. Note that when the temperature gets near the freezing point of your dilution, the solar water heating system will stop working, but you will still be protected from bursting pipes to a much lower temperature. The leanest mix of propylene glycol you should consider in the mix is 20 percent glycol. The richest mix of propylene glycol you should consider in the mix is 60 percent glycol. Remember to always use the leanest percentage of glycol that you can get away with and still have the protection you need for your climate.

Figure 4.8: *Sample glycol freeze protection table*

Temperature C	Temperature F	Percent Glycol Concentration Required: For Freeze Protection Volume%	Percent Glycol Concentration Required: For Burst Protection Volume%
-7	20	18	12
-12	10	29	20
-18	0	36	24
-23	-10	42	28
-29	-20	46	30
-34	-30	50	33
-40	-40	54	35
-46	-50	57	36
-51	-60	60	37

It is important to note that glycol is slippery and harder to pump than water because of its relative viscosity. It is important to note that most pump curves relate to pumping plain water. Because glycol mixes are harder to pump than pure water, it is important to consider the viscosity of glycol when sizing your pump.

Propylene glycol fluid in a solar water heater will deteriorate over time and will eventually wear out. The solar fluid has to be checked periodically. Details are given for this procedure in the maintenance section of Chapter 9. On average, a high-quality and high-temperature glycol solar fluid will last 15 to 20 years. If the fluid is subjected to abnormally high temperatures or long periods of stagnation (no flow), its life will be reduced. If either of these situations occur, the fluid should be checked more often. When glycols deteriorate, they become more acidic, which is harmful to the system, and their freeze protection capacity is diminished. The freeze point of glycol/water solar fluid is checked with a refractometer and the acidity is usually checked with litmus paper.

Other System Components

Expansion Tanks

Liquids expand when they are heated; this expansion can be substantial. Unlike gases, liquids cannot be compressed. In a solar water heating system, the solar fluid can experience dramatic temperature changes over the course of a single day. The expansion and contraction of the solar fluid over this range of temperatures must be compensated for, or the system will burst. An expansion tank is the component that does that compensating.

Expansion tanks contain a membrane or bladder that separates a volume of air in a chamber that is open to the main circuit. As the fluid is heated and expands, pressure builds up in the system. As the pressure increases, the air in the expansion tank compresses and makes room for the expanded liquid. Commercially available expansion tanks have a bladder separating the liquid from the air. The air pressure can be adjusted in the air chamber via a Schrader valve on the bottom of the tank. The air pressure should be set to three pounds per square inch below the pressure of the system at 60°F fluid temperature. This can be measured with a regular tire gauge. See the section on pressurized antifreeze systems in Chapter 5 for pressure suggestions. A common cause of solar water heating system failure is related to undersized expansion tanks. When the tank is too small, excessive pressure can build up in the system and can damage components. Sizing guidelines will be

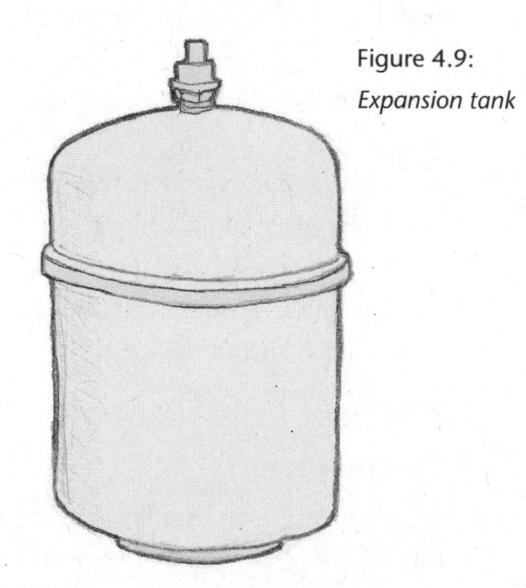

Figure 4.9: *Expansion tank*

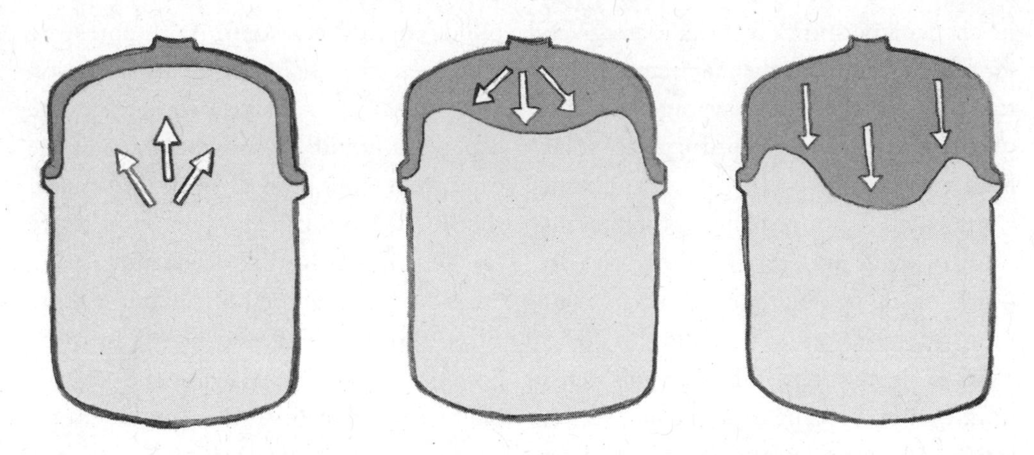

Figure 4.10: *Expansion tank process*

covered in Chapter 7. It is also important to use an expansion tank that has a bladder that is compatible with glycol.

Attempts have been made to use an expansion chamber in lieu of an expansion tank with a bladder. An expansion chamber is essentially a long tube or a sealed tank that is isolated from the rest of the system during system charging and is later opened once the system is under pressure, effectively leaving a captured bubble in the system. This method eliminates a rubber bladder, which deteriorates over time. However, almost all attempts at this method that we have seen in the field have eventually saturated the chamber with fluid, causing the air to be released throughout the system.

Thermometers

It is interesting to have some thermometers installed at various locations on the solar energy system. They help you monitor the performance of the system. They are not necessary for the operation of the system; however, we consider them to be essential for monitoring, inspection and troubleshooting, and we put several on every system we install. A temperature gauge is the best means for a system owner to determine whether the system is functioning properly. There are two types of thermometers to consider: mechanical or electronic. Mechanical thermometers typically mount in a well or chamber placed within the pipe that carries the liquid you want to monitor. Note that you have to plan ahead and locate the wells in the proper place as the circuit is being plumbed. Well thermometers are accurate and relatively inexpensive. Electronic thermometers use sensors that are fixed to the pipe where you want to monitor the temperature.

A well is not required, so no prep work needs to be done. An advantage of electronic thermometers is that you can often locate the temperature display some distance from the site of the sensor. People often locate the display in an area that they frequent, so it is easy to see what is going on without having to be present at the site of the sensor. Some equipment allows a single display that shows the temperatures at several different locations. There are even more sophisticated temperature monitors that will interface with your computer, so you can do accurate data logging and energy metering.

The most common place for thermometers is on either side of the heat exchanger on the solar loop. This placement allows you to see the temperature of the incoming solar fluid and the temperature drop across the heat exchanger. It is also a good idea to put a thermometer on the solar storage tank, near the top, to monitor the temperature there.

Pressure Gauge

All closed-loop pressurized systems require a pressure gauge. The gauge can be placed anywhere on the solar loop, but it should be in close proximity to the charging ports. (See "Charging a Pressurized System" in Chapter 9). Pressure gauges are inexpensive, so get a good one. You can get a combination temperature/pressure gauge.

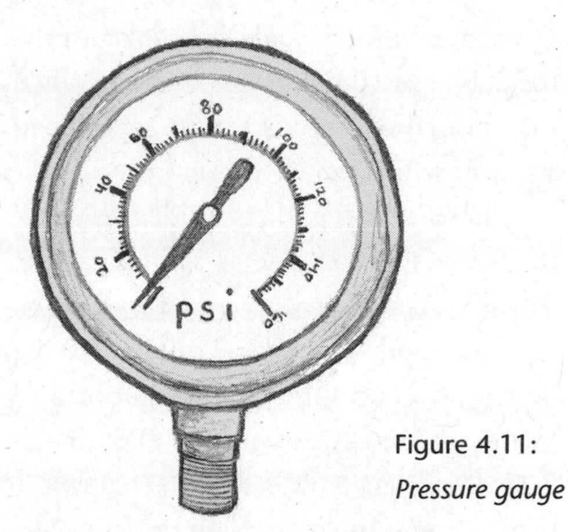

Figure 4.11:
Pressure gauge

These work fine, but again, get a good one as the cheap ones fall apart.

Check Valves

Check valves allow the fluid passing through them to flow in one direction only. An arrow on the valve body indicates the direction of flow. Two kinds of check valves are commonly used in solar water heating systems: spring check valves and swing check valves. Both kinds have their advantages and disadvantages. Bob has used swing check valves exclusively for more than 25 years with no failures.

Swing check valves control the direction of flow by using a swinging door inside the valve. When the fluid is flowing in the direction of the arrow, the moving fluid pushes the door open and flows through the doorway. When fluid flows in the other direction, the fluid pushes the door shut, stopping the flow. These valves are typically made of bronze, so the metal is fairly hard and very durable. When the door closes, it rests against a machined surface, which is very flat and even.

There are no gaskets inside the valve; the seal is simply metal against metal. These valves can be located on horizontal pipe runs and vertical pipe runs where the fluid flow is upwards. This is important! Swing check valves will not work if the flow direction arrow is pointing anywhere below horizontal (pointing down at any angle).

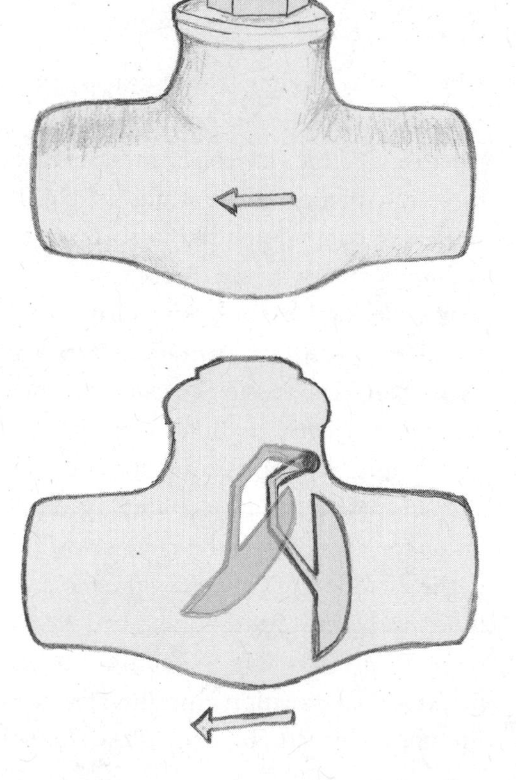

Figure 4.12: *Swing check valve*

Figure 4.13: *Swing check valve process*

Spring check valves use a spring that holds a door or a ball closed. Fluid flowing through the valve in the direction of the arrow will compress the spring and allow the door to open. These valves can be positioned in any direction. They do not rely on gravity to close the door when flow stops. Most spring check valves are produced to operate in systems that use powerful pumps or relatively high flow conditions, so the springs are often quite stiff. This can cause a serious restriction when small pumps are used, as is often the case in solar energy systems. Some specialty spring check valves use modified soft springs that open more easily. Some experienced installers modify the springs in commercial spring check valves by opening up the valves and cutting the springs down so they open more easily. Another tactic is to remove the spring from the valve completely, but the check valve must be installed in a vertical pipe where the fluid moves in an upward direction.

Some installers have used a motorized valve to close the system down. Although this can technically work, it is a bad idea because it adds an unnecessary level of complication and expense. It also introduces another part that can potentially wear out or fail in a system. Keep it simple!

Many pumps now have a small plastic check valve inserted into either the suction or discharge port. Do not rely on this as the check valve in your system. It is not sturdy enough to last over time. You should have only one check valve on the solar loop to prevent isolating some portion of the system, so you will want to ensure that you remove any check valves that come with your pump(s).

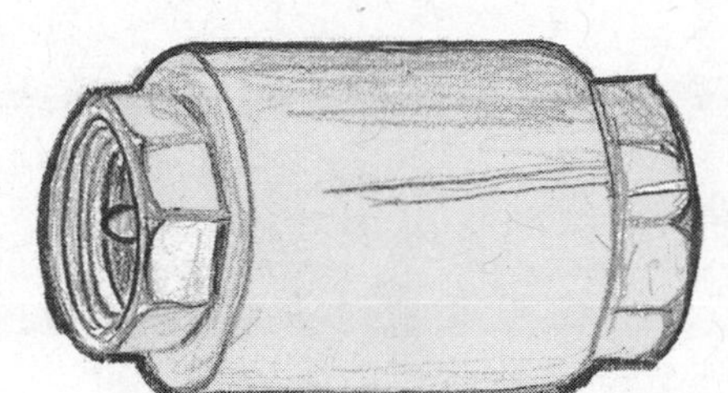

Figure 4.14: *Spring check valve*

Differential Temperature Controller

A differential temperature controller (DTC) is an electronic device that can compare the temperatures of two remote locations and has a logic component that can turn a relay on and off at appropriate times. It has a built-in microprocessor that can turn on a pump or pumps when the temperature is warmer in one location than in the other. This controller uses electronic sensors to measure the temperatures at each location. It includes a relay switch that turns the pump on or off. Most controllers are adjustable, so you can set the temperature differential that the logic will use to operate the pump. Some controllers may also have a high-limit feature that turns off the pump when the storage temperature reaches a preset temperature or a recirculation feature that circulates warm fluid from the solar storage tank when collectors near freezing temperatures. On a closed-loop antifreeze system, these two features should be disabled.

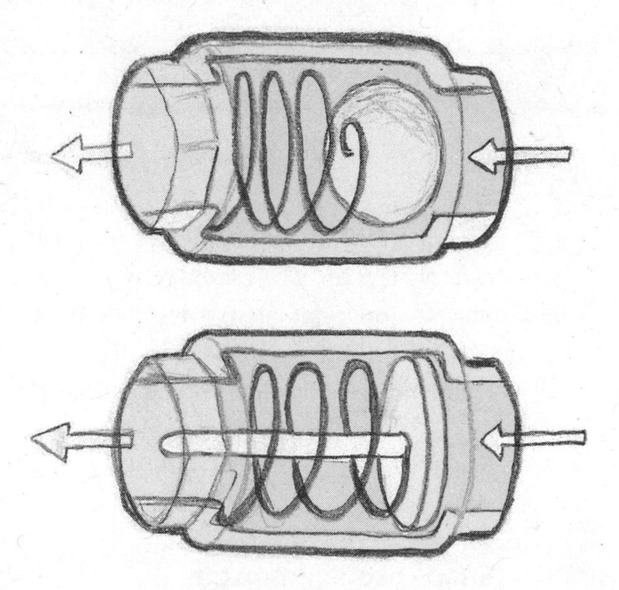

Figure 4.15: *Spring check valve process*

Many recent controllers have upgrade options that may include variable-speed pump control, WI-FI compatibility for remote monitoring, data logging, Btu calculations, summer cooling and vacation settings, and irradiance measurements in addition to many others. One of our favorite new controller innovations is a dual-relay controller made especially for pressurized systems. With this controller, when the storage tank reaches high limit,

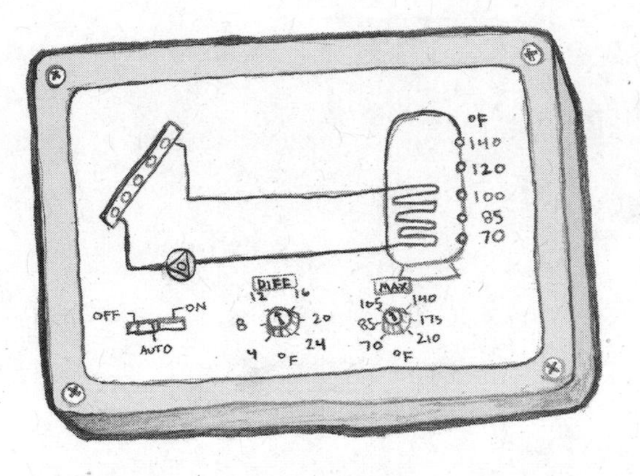

Figure 4.16: *Differential temperature controller*

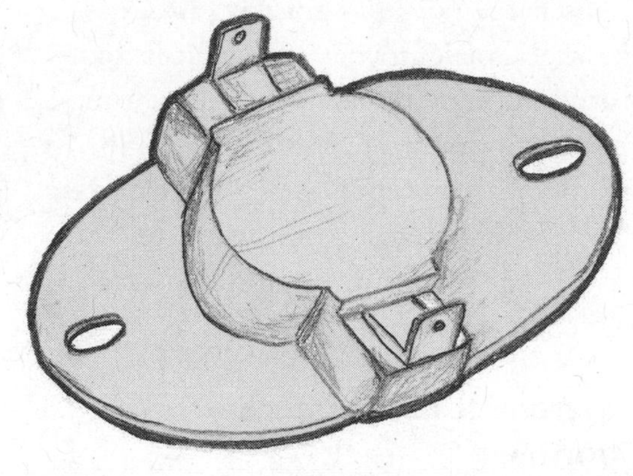

Figure 4.17: *Snap disc thermostat*

a second relay engages while the main circulator stays on. The second relay can turn on a secondary pump or valve that diverts the solar fluid to a diversion or shunt load, thereby cooling the fluid but keeping the fluid circulating at the same time, which will extend the life of the solar fluid dramatically.

Snap Disc Thermostats

A snap disc thermostat is an inexpensive switch that turns a circuit on at a preset temperature. Most snap disc thermostats are not adjustable. These thermostats are not very accurate or durable and should not be used in a solar water heating system.

Aquastat

An aquastat is a thermally activated switch that turns on and off at different temperatures. It is usually adjustable to meet the system's needs. It is a more accurate and more expensive type of thermostat switch that measures temperature that allows for advanced system functions. An aquastat is better than a snap disc switch, because a snap disc can be set for one temperature only. With the aquastat you can set the temperature at which it will go on and off, and also set the differential. It is not used to control the main flow through a solar loop, but to serve a particular function within the system. Part of the aquastat is a sensor that is like a thermal couple. A thermal couple is a temperature sensor that creates its own millivoltage; the voltage it creates is determined by the temperature of the sensor, or the bulb. There are two types. One has the sensor built in to a strap that fastens on to the pipe, keeping the sensor in contact with the pipe. The other, called a remote bulb thermostat,

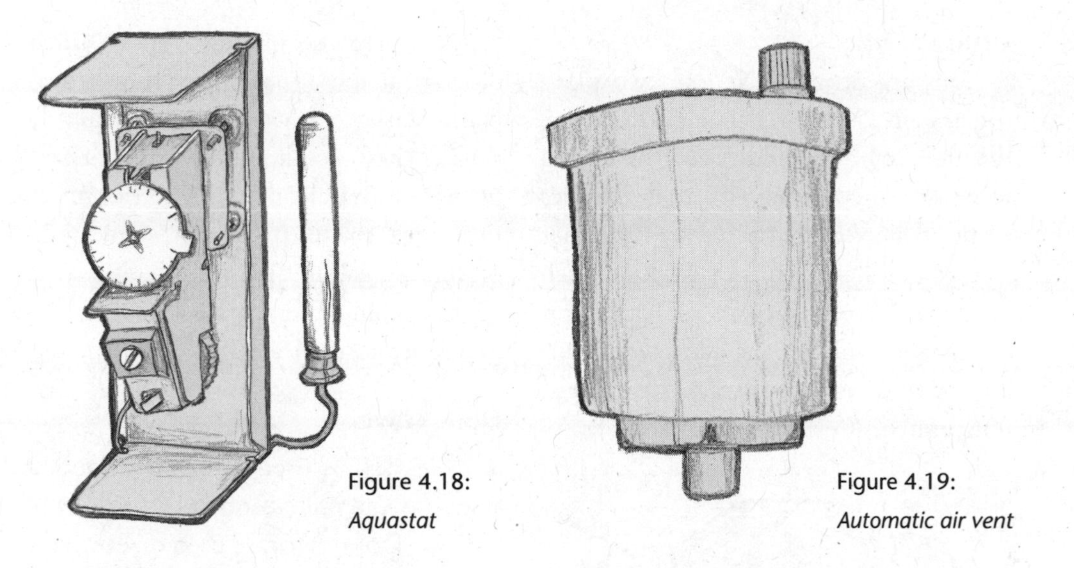

Figure 4.18:
Aquastat

Figure 4.19:
Automatic air vent

has a small copper tube that connects the sensor to the aquastat.

Automatic Air Vent

Automatic air vents are valves that allow air to escape from a circuit that is full of fluid. On the solar loop they are placed at the highest point in the system where air will accumulate. These valves are especially helpful in traditional hydronic heating systems, where fresh water is continually added to the system. They can be an important component in a solar water heating system when placed on the hot water side of the heat exchanger, where fresh water is always present. Note that fresh water always contains a small percentage of dissolved oxygen and that oxygen starts to form bubbles as the water is heated in the heat exchanger. If the bubbles are allowed to add up to a big bubble, this can impede the flow of water through the heat exchanger, especially if the heat exchanger is set up to thermosiphon on the water side.

Automatic air vents come in a wide range of designs and qualities. The cast brass units are best, and the larger units are better as well. Be sure to select a valve that has a high-temperature float. When you are installing an air vent on the solar loop of a closed-loop antifreeze system, we recommend that you isolate the air vent with a ball valve. Once all of the air has been released from the system, close the ball valve. Since the solar loop is a sealed system with a fixed amount of fluid and dissolved oxygen, it is not necessary for the vent to be open to the system except when recharging.

Unions

Unions are fittings that join two pipes together in a plumbing circuit. They allow the pipes to be taken apart without having to be cut. They are used on all the nipples of the collectors and are used on either side of critical parts in the solar loop, such as the circulating pumps and the heat exchanger/tank fittings. The best unions are made of solid brass and have high-temperature O-rings or gaskets.

There are also specialty unions called dielectric unions. These have a rubber-and-plastic gasket and washer that electrically isolates the two sides of the fitting. These fittings are usually placed where connections are made to the storage tank, preventing galvanic reactions between the copper piping and the steel tank. It is a good idea to use them.

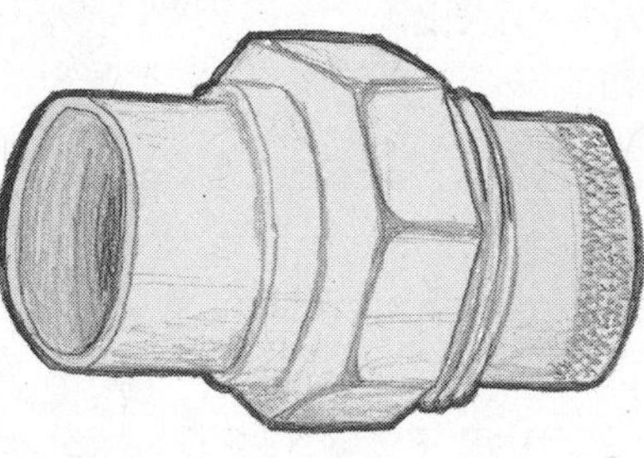

Figure 4.20: *Dielectric union*

Figure 4.21: *Dole valve*

Dole Valve

A Dole valve is used to protect flooded systems from mild freezing conditions.

Two Dole valves are used on each collector array, one on the bottom manifold end if flat plate collectors are used. They open a drain as the temperature drops to 44°F, and this allows fresh warm water to flow through the array, protecting the collectors. These valves are used only when freezing conditions are a very rare occurrence. I do not recommend using them, for a number of reasons. First, I never recommend flooded systems except ICS systems. Second, if you are in a climate that has any possibility of experiencing freezing conditions, you should choose either a pressurized or a drainback system. However, they are often installed on ICS systems simply as a precaution.

Ball Valves and Gate Valves

Ball valves and gate valves are used to stop the flow of fluid in a piping circuit.

The traditional gate valve has a spin handle that operates a gate that slides up and down within the valve. Gate valves always create resistance within the circuit and are intended to be operated only in the fully open or fully closed position. They should never be used in a solar energy system.

Ball valves have a lever attached to a ball that rotates within the valve. The ball has a hole running through the middle of it. When the ball is rotated to the open position, fluid can pass through the hole in the ball and flow can proceed. When the ball is rotated to the closed position, the hole is perpendicular to the flow and flow is stopped. Ball valves can be set in any position to regulate the flow through the circuit. Always use "full flow" or "full port" ball valves.

Figure 4.22: *Gate valve*

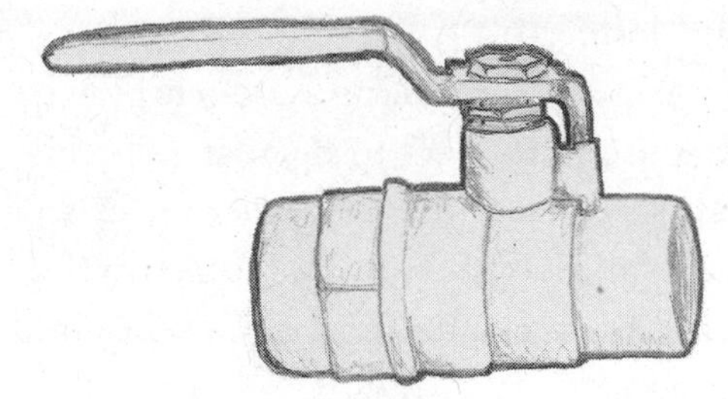

Figure 4.23: *Two-port ball valve*

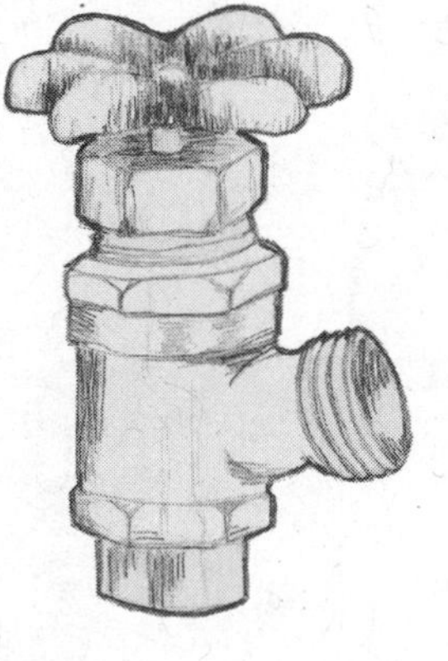

Figure 4.24: *Drain valve*

Drain Valve

Drain valves do exactly what their name implies. They allow you to drain fluid from the system. Most will be of the same quality. They are used in every type of solar water heating system and are a very common plumbing implement. The drain can be shut off with either a gate or a ball valve. The ball valve types are much better at standing up to higher pressures. We recommend only the ball valve type. In either case, be sure to cap off drain valves when not in use.

Relief Valves

A pressure-relief valve protects circuits from excessive pressure buildup. It has a gasket attached to the end of a plunger that has a heavy spring holding the valve closed. When pressure in the system exceeds the rating of the valve, the spring compresses, the valve opens, and pressure is released. For a pressure-relief valve, it is important to attach a drain pipe to the valve outlet and to run the pipe so that it terminates in an out-of-the-way location, so if the valve opens, hot pressurized water will not spray on anybody. Most codes require them to terminate within 6" of the floor.

You can purchase pressure-relief valves that are factory set to open at a specific pressure; other pressure-relief valves are field adjustable. This is a required valve for every pressurized solar water heating system.

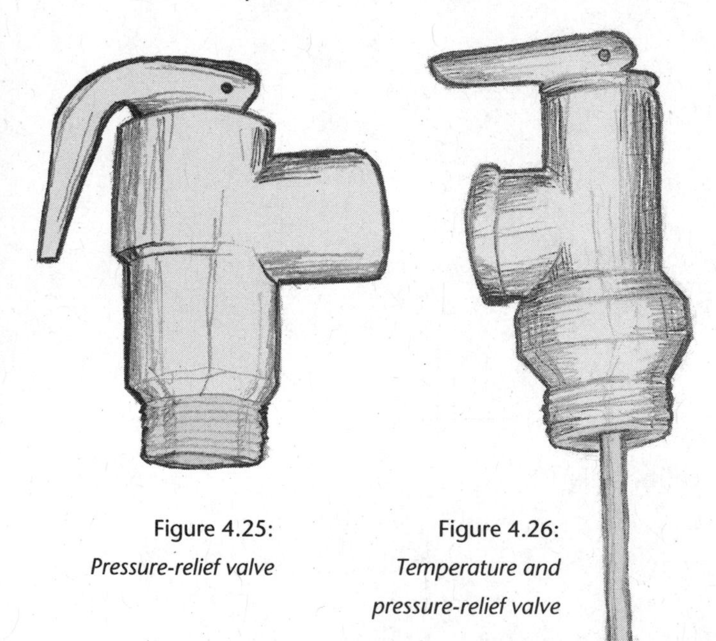

Figure 4.25: *Pressure-relief valve*

Figure 4.26: *Temperature and pressure-relief valve*

Another similar valve is a temperature and pressure-relief valve. These valves open either when the pressure in the system exceeds the preset pressure or when the temperature in the system exceeds its preset temperature. These valves are designed to be installed on conventional gas or electric water heaters to protect them from overheating or over-pressurization. Most codes require them on any tank that is connected to a heat source. These valves should never be used on the piping in the solar loop because the high temperatures of the solar fluid may set them off.

Mixing Valves

A thermostatic mixing valve (TMV) combines a mix of hot and cold water so the finished product is at the desired temperature. Solar water heating systems often heat water to dangerously high temperatures. These valves should be installed on **every** solar water heating system to protect the hot water users from being scalded. Because this is an important part of the system, make sure you purchase a good quality unit.

A mixing valve has three ports: hot, cold and mix. There is also a handle, which is turned to set the desired temperature of the mix. If the flow of hot liquid through the valve is below the set temperature, the fluid flows straight through the valve and no mixing takes place. If the fluid entering the valve is hotter than the set temperature, some cold fluid is added to the hot fluid to cool it down to the desired temperature. When the water in the solar storage tank is too hot to be safe, the tempering valve is used to lower that temperature. It is important to always install a check valve on the cold line going into the tempering valve to prevent heat from bleeding into the cold water line. Many mixing valves have built-in check valves to prevent thermosiphoning and heat loss.

Thermostatic mixing valves should not be confused with tempering valves. Although the terms are commonly used synonymously, the equipment is quite different and they are not interchangeable. A tempering valve is controlled mechanically instead of thermostatically. It mixes hot and cold fluid at a pre-set volume or ratio, not at a pre-set temperature. Because the temperature in the solar storage tank may vary, the end result may be a domestic hot water temperature that varies during use. You or your clients will be unhappy and it may result in injury.

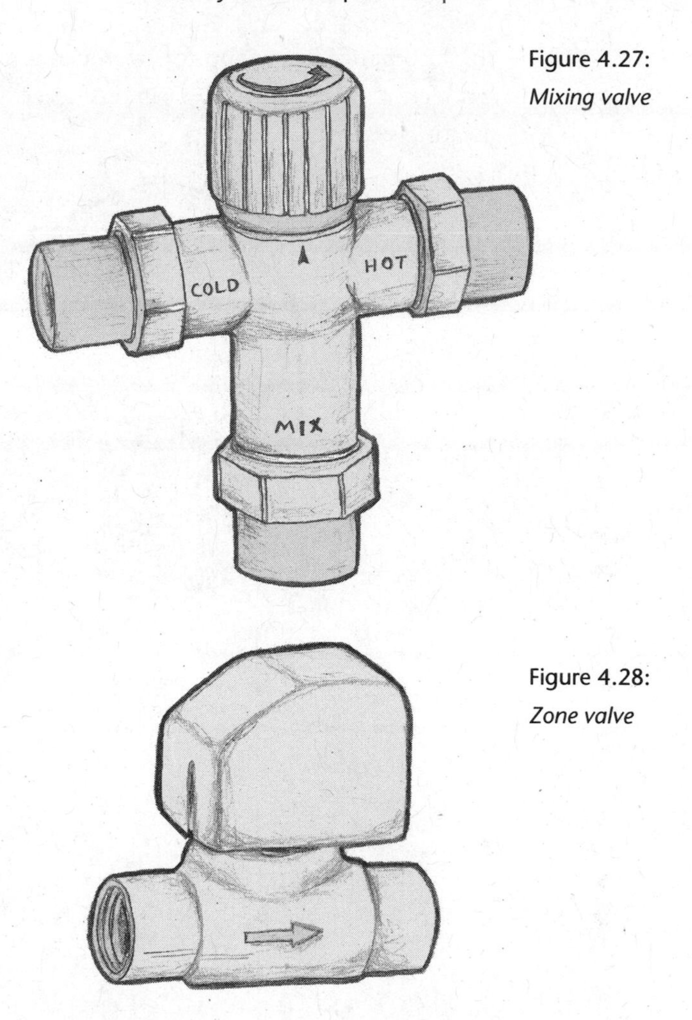

Figure 4.27: *Mixing valve*

Figure 4.28: *Zone valve*

Motorized Valves

Often called zone valves, motorized valves turn a circuit on or off by using a motor to operate the valve. They are typically used in hydronic heating systems to control the flow of heat to various zones within a building. Some motorized valves operate on 120-volt AC; others operate on 24-volt AC, the typical operating voltage for heating controls. Note that there is always a direction of flow indicated on the valve housing. Because of the electronics involved, it is not recommended to locate a zone valve on the hot supply side of the solar loop. They are occasionally used to control an automatic shunt loop for overheating protection on closed-loop antifreeze space heating systems.

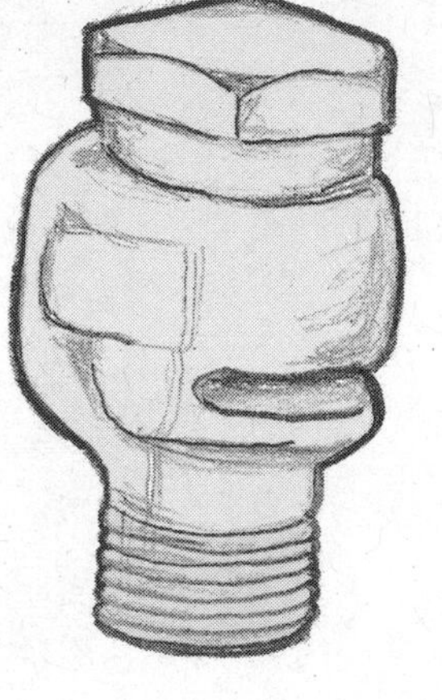

Figure 4.29: *Vacuum breaker*

Vacuum Breaker

Vacuum breaker valves are used to let air into a system when the system is depressurized. They contain a plunger with a gasket on the end. When the system is under pressure, the plunger is forced against a seat, sealing the system. When the pressure is lost, the plunger drops away from the seat, and air is allowed into the system through the valve. Typically found in solar swimming pool systems and some draindown systems, these valves facilitate fast system draining by allowing air to enter the system. They can also be used on the top of a fiberglass water heating or storage tank, where they help prevent the tank from collapsing when being drained.

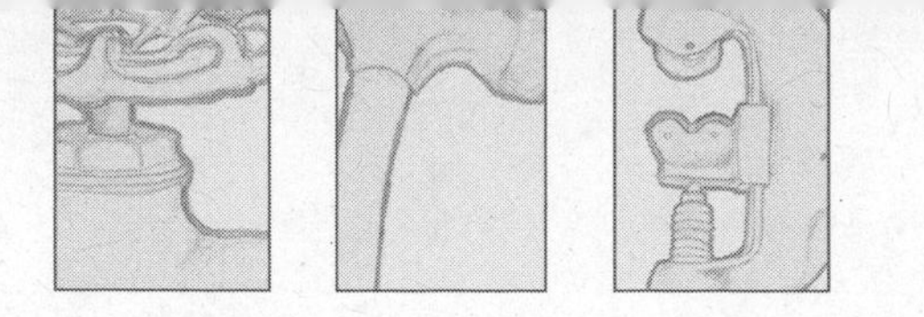

5

SOLAR WATER HEATING SYSTEMS

Introduction

BEFORE WE START detailing the various system designs for solar water heaters, it is best to clarify some terms that are used to classify and define the systems. We will also go over some basic principles that are essential for understanding the system designs.

Be sure to read all of this chapter to fully understand solar water heating systems. Each section explains a particular type of solar energy system, telling how it works and the basic components it uses. Another reason to read each section is that early sections provide definitions of concepts that are used in following sections.

Recommendations

Although a number of systems have been used in the past, there are really only three domestic water heating systems that we typically recommend: ICS, drainback and pressurized. These three are the only systems that have proven their reliability over the past 30 years. Some of the systems described in the following pages are identified solely for the purposes of education and repair. If you are a solar professional, you need to be prepared if you are called on to service one of these systems.

Basic Terms

Open-Loop, or Direct, Systems

In an open-loop, or direct, type of solar water heating system, the domestic water (the water we use in our home) actually enters the solar collector, where it is heated.

Closed-Loop, or Indirect, Systems

In a closed-loop, or indirect, type of system, a heat transfer fluid (also called solar fluid) is heated in the collectors and then circulated to a heat exchanger, where the solar heat is transferred from the solar fluid to the domestic water. Closed-loop systems can use flat plate, evacuated tube or concentrating collectors. Flat plate collectors are the most common.

Passive Systems

Passive solar water heaters circulate either domestic water or a heat transfer fluid through the system without the use of pumps. Typically, passive systems have few (if any) moving parts and require no external energy to operate. Being very simple, they tend to be very reliable and easy to maintain. Because of their simplicity, they also tend to be the least expensive system to choose from. But because of their simplicity, they may be vulnerable to problems that other systems can overcome. These systems may under-perform because optimal flow-rates may not be achieved on a regular basis. Care must be taken when installing the piping to assure efficient circulation.

Active Systems

All active solar water heaters utilize pumps to circulate fluids throughout the system. Some form of external power is required for operation.

Two-Tank Systems

All systems need some type of storage tank to store the water that is heated by the sun. Most systems use a storage tank for the solar-heated water and a backup heater that can be either a tank type of heater or an on-demand (tankless) water heater. These systems are called two-tank systems. The tanks are configured in series, with all of the domestic water first entering into the solar storage tank, where it is heated by solar energy. This tank feeds into the backup water heater, which allows the solar storage tank to preheat the water. If it has been a sunny day and the solar storage tank is sufficiently hot, the backup water heater will not have to run. However, if it has been cloudy and the solar storage isn't hot enough, the backup heating source will need to turn on to bring the water up to the desired temperature as water is introduced to that tank. However, even if the solar storage hasn't got the water quite up to the right temperature, it will have at least got part way there, thereby minimizing the amount of energy the backup will need to consume.

One-Tank Systems

Some systems use the solar storage tank as the backup heating system as well by including an electric heating element or a gas burner to heat the water when there is insufficient solar energy available to bring the water up to the desired temperature. These systems are called one-tank systems. Optimally, the single tank will be as large as both the solar storage tank and the backup water heater in a two-tank system.

One concern with one-tank systems is that if hot water is used during the evenings and depletes the storage tank of hot water, the electric or gas heat source will have to turn on to heat the tank to the set point temperature. Then when the sun comes out the next day, the tank is already warm and the efficiency of the solar water heater is greatly diminished. This is especially true in climates where the sun is less than optimal. For most climates we have found two-tank systems to be the best choice for optimal performance. In warm and sunny climates, where the solar resource is reliable on a daily basis, one-tank systems are more acceptable. All of the illustrations of systems in this book are of two-tank systems.

Basic Principles

Heat Transfer

Since all of the following systems deal with the movement of solar heat to our desired demand, it is important to review the principles of heat transfer. When a fluid is heated, its molecules expand. As molecules expand, they become lighter and rise above any cooler (denser) surrounding molecules. If we observe a tank of water with some water in it that is hotter than the rest, the hottest water will rise to the top of the tank. This phenomenon is called stratification. This principle will be used to our advantage in most solar water heating systems.

It is important to note that heat doesn't rise; hot fluids rise. Heat can travel in any direction, but will always transfer from the hot source to the cold source. The rate of heat transfer will increase as the temperature differential between the two substances increases.

Freezing Water

Compounds contract as they cool. As they contract, they become denser and heavier. Eventually, a liquid can become so dense that it becomes a solid. This is the opposite of what happens as compounds are heated. Water is unusual among compounds in that most compounds are densest in their solid state, whereas water is densest just before it freezes (becomes a solid) and then expands again as it gets colder. If water acted like most other compounds, ice would sink instead of floating when surrounded by liquid water. Life as we know it would be vastly different, if it would be here at all, if this happened. Just think — if ice sank instead of floating, as lakes froze, the ice would sink and crush all the fish and plants. All our lakes, rivers and oceans would freeze solid and probably never thaw out. This quirk of physics, where ice floats on water, allows life as we know it to exist.

This expansion is a very powerful force. Freezing water has shaped the earth by breaking rocks into soil. Water gets into cracks in rocks, and when it freezes and expands it can literally explode a rock apart. This can be a bad thing in a solar energy system. If water in a pipe freezes, it expands. Most of the time, this expansion exceeds the limits of what a pipe can take, and the pipe breaks.

Water Quality

In most locations, our water supply is considered "hard." Hard water is water that contains dissolved minerals. Most of us are familiar with water softeners: they remove these minerals.

When hard water is heated under certain conditions, these dissolved minerals can precipitate out of the water and solidify. People who have hard water often experience problems with conventional water heaters. Electric water heaters tend to wear out an excessive number of heating elements. A gas water heater tends to build up mineral deposits on the bottom of the tank (where it gets the hottest), and this sometimes causes the heater to make various noises when the burner is ignited. Hard water is also one of the main causes of premature tank failure. This same thing can happen in a solar collector. Mineral buildup inside a collector will decrease collector efficiency by depositing a layer of minerals on the inside of the collector. This mineral buildup acts like a layer of insulation on the inner surface of the waterway and slows the transfer of solar heat to the water. In collectors with small waterways, this mineral deposit can actually plug the pipes and render the collector useless.

Pressurized Antifreeze Systems

The precise description of this system would be a pressurized closed-loop antifreeze system. We will call this type of system pressurized for short. It is the most versatile of all solar heating systems and the most widely distributed type worldwide. The collector arrays can be mounted in almost any imaginable way and located at considerable horizontal distances from the exchanger.

This is an excellent choice for all climates when there is a reliable and consistent water heating load every day. It is the only fail-safe system for climates that experience freezing conditions and should be the only choice for climates that experience prolonged or severe cold weather or for climates that experience heavy snowfalls. It can also be used in all climates where hard water conditions are encountered.

Pressurized systems are indirect active systems. With the exception of pressurized thermosiphon systems, pressurized systems always use a pump or pumps to circulate fluids through the system. These systems commonly use flat plate or evacuated tube collectors that heat a solar fluid, usually a high-temperature propylene glycol-water mixture, which is circulated through them. Two insulated copper pipes connect the collectors to a heat exchanger. A relatively small circulating pump is installed on the pipe that feeds the collector array, along with an expansion tank, drain and fill valves, a check valve, a pressure-relief valve and a pressure gauge.

A loop of piping starts at the heat exchanger, travels through the collector array, and then completes the loop, traveling back to the heat exchanger. This loop of piping is completely sealed and is kept completely full of the solar fluid at all times. This type of system can be controlled and powered by a PV panel and a DC pump (with or without a controller) or can use a differential temperature controller

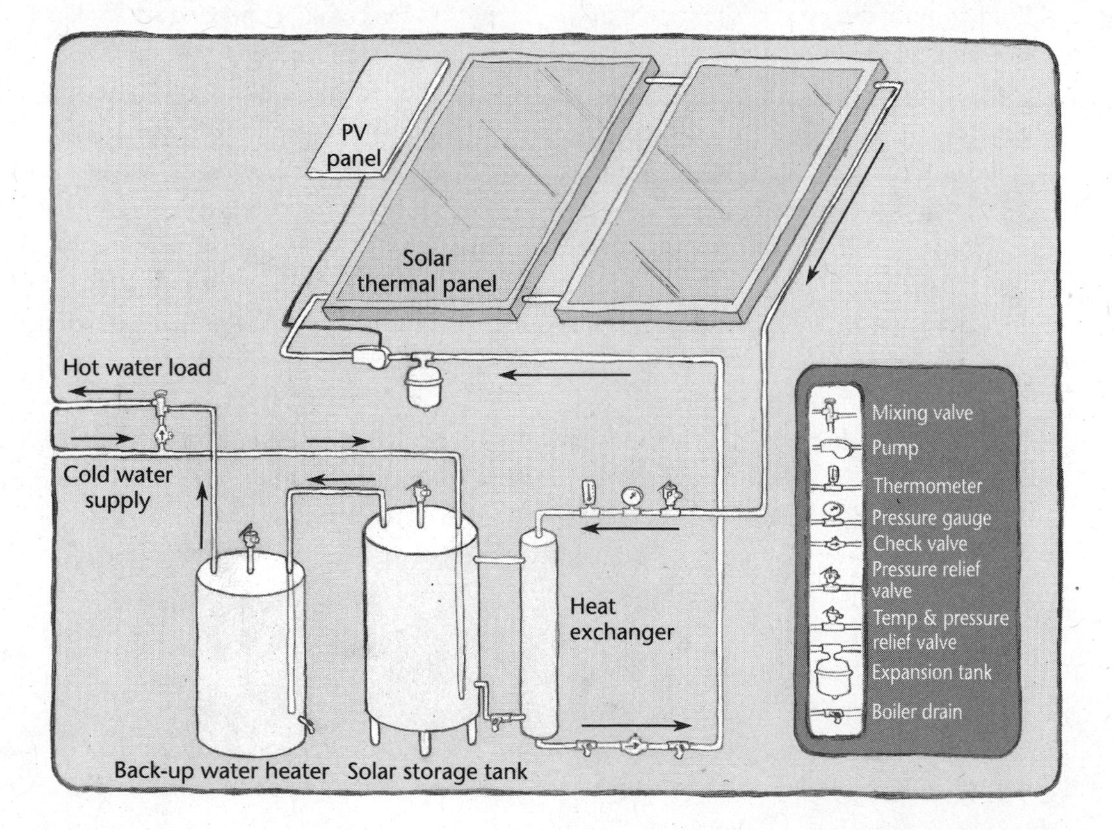

Figure 5.1: *PV-powered pressurized antifreeze system*

operating a 120-volt AC pump. When the pump starts, the solar fluid is circulated through the collector array, through the hot supply pipe to the heat exchanger, and then back to the collectors through the return line. When the pump stops, the solar fluid simply stops moving within the closed loop. Because the solar fluid stays in the whole closed loop at all times, there is always fluid in the collector array and the piping, which travels through unconditioned space. The solar fluid must be able to not only act as a heat transfer fluid, it must also protect the system from freezing. Please study the earlier section that describes the characteristics of antifreeze. Methods for transferring the heat from the solar fluid to the domestic water will be described below.

The major limitation of pressurized systems is the tendency of the solar fluid to degrade under high temperatures. The best antifreeze solution today is high-temperature-formulated propylene glycol (described in Chapter 4). This fluid can eventually break down and form a corrosive solution that can harm system components. High temperatures degrade the fluid, and the rate of degradation is directly proportional to the intensity of the overheating, so the hotter it gets, the more quickly the fluid will degrade. All overheating scenarios can be avoided, so this should not be considered a fatal flaw. The best way to reduce overheating is to make sure the system circulates fluid whenever it is sunny out. It's very important, if you have this type of system, to check the condition of the solar fluid regularly. Propylene glycol solar fluid solutions need to be changed periodically. Think of it like changing the oil in your car or truck, except that you have to do it only once every 10 to 15 years under normal conditions.

Pressurized systems are used for both residential and commercial water heating systems and space heating systems. Most manufacturers offer pressurized system kits that are pre-engineered to take the guesswork out of designing a system. They work great for the majority of residential or small system installations. Because most residential solar water heating systems are the same size and installed the same way, these kits are fine for those situations. I highly recommend these kits if purchased from a reputable manufacturer. System sizing is covered in Chapter 7, but you should know that there are only a handful of system sizes that will satisfy the majority of domestic water heating demands, so system sizes and system components are fairly uniform.

To make sure that pressurized systems function properly, a few rules must be adhered to:

- The hot supply pipe between the collectors and the heat exchanger must always be made of copper or stainless steel and must be insulated with high-temperature pipe insulation rated to at least 250°F because under rare occasions it can get very hot for short periods of time. Copper or stainless steel are the only practical materials that can withstand this heat.

- In residential solar water heating systems, the return line from the exchanger to the collectors should also be copper or stainless steel, but standard high-quality pipe insulation rated to at least 180°F can be used on that line.
- In large solar space heating systems, the return line can be made of Pex tubing, except for the final 10 feet, where it attaches to the collectors.
- Collector arrays should be mounted with a slight slope toward the return inlet where the solar fluid enters the array, and all piping should slope slightly toward the heat exchanger. For arrays that are remote, a drain should be placed at the inlet where the solar fluid enters the array to facilitate drainage.
- All horizontal pipe runs should be supported every 5 to 6 feet, and all vertical runs should be supported every ten feet. Corrugated stainless steel piping should be supported every 3 feet on non-vertical pipe runs.

Large space heating systems and large commercial systems are often pressurized. These systems can be very large and may be the only option for any commercial application where the collectors cannot be located above the storage tank and heat exchanger. A complete discussion of these types of systems is included in Chapter 6. The world's largest flat plate solar water heating system for many years was in Green Bay, Wisconsin. It used more than 5,250 4' × 8' flat plate collectors (157,689 square feet of collector area). This was a pressurized solar water heating system, and it performed extremely well. That system produced more than 37,500 million Btus annually. This system was taken out of service when natural gas prices reached very low prices during the late 1980s.

Large space heating systems, seasonal systems and systems that may experience periodic idle times with no load present will require a heat-diversion load, sometimes called the shunt load. This shunt diversion load helps keep the solar fluid from overheating. A shunt loop is typically a buried length of uninsulated pipe or a radiator located outside, preferably in a cool or windy spot. Exposed fin-tube is very popular. Outdoor hot tubs are also a common shunt load. Designs and sizing parameters for diversion loads will be discussed further in Chapter 6.

Thermosiphoning

Because fluid remains in the solar loop at all times, a method to prevent thermosiphoning must be included in the solar loop. Thermosiphoning can happen when the collectors are colder than the storage tank, but only if the collectors are located above the storage tank, as with roof-mounted arrays. Heat from the storage tank and heat exchanger could rise up the supply pipe to the top of the array, and cold heavy solar fluid could drop down the return pipe to the heat exchanger from the bottom of the array. Under this scenario the fluid is circulating in the opposite direction than when it is heating. A check valve will stop this from happening. If a check valve is used in the system, it is typically located between the two boiler drains that are used to charge the system. The check valve will also make sure the fluid flows correctly when charging the system. (Check valves were more fully described in Chapter 4.) They are required in the solar loop only when the collectors are above the heat exchanger and storage tank. If the collectors are below, reverse thermosiphoning will not take place. If no check valve is used in the closed loop, then a fullport ball valve must be installed between the drains to facilitate charging. There are occasions, however, when reverse thermosiphoning is a good thing. For instance, if you are going on an extended vacation (longer than two weeks) you may want to keep your tank cool to prevent overheating and speeding the degradation of the solar fluid. You can install a vacation bypass around the check valve to allow for reverse thermosiphoning to occur. This method is most effective if flat plate collectors are used but would not be effective if the collectors were located below the storage tank.

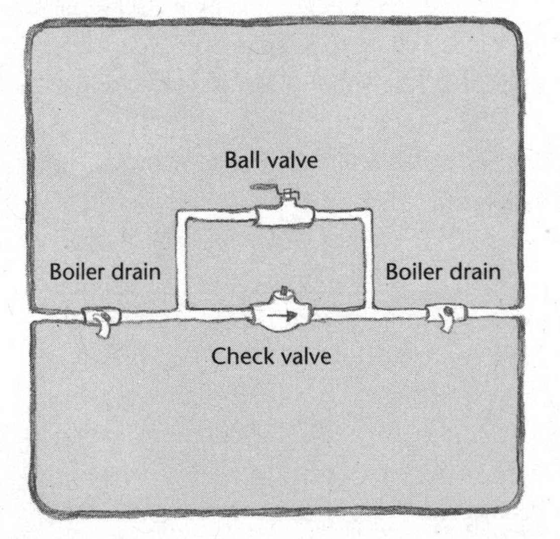

Figure 5.2: *Vacation bypass*

Pressure, Expansion Tanks and Pumps

The solar fluid within the solar loop is put under pressure when the system is charged. In cold climates the pressure is typically set at a minimum of 35 pounds per square inch at 60°F fluid temperature, but it can be less in hot climates. In cold climates the solar fluid contracts significantly. If there is not enough pressure in the system to compensate for this loss of volume, the pressure in the solar loop can actually become negative. When that happens, air is sucked into the solar loop through valves and other components, and circulation is inhibited and can actually stop. This is commonly called vapor lock and is very bad.

When the fluid is heated, it expands. As it expands the pressure within the closed loop rises. Eventually, the pressure could rise enough to burst the piping. To prevent such a catastrophic event, an expansion tank is installed on the closed loop. (Expansion tanks are fully described in the Chapter 4.) When the fluid cools during the night, especially during the winter in cold climates, the pressure will drop because the fluid contracts. The expansion tank will provide some compensation for this situation too. It is not unusual to see the pressure in the solar loop fluctuate between 10 and 45 pounds per square inch seasonally, and sometimes daily in cold climates. Set the pressure of the expansion tank at three to five pounds per square inch below the pressure in the system at 60°F, as detailed in the Chapter 4.

The best location for the expansion tank within the closed loop is directly before the circulating pump in relation to fluid flow. This maximizes flow by compensating for the negative pressure that is created on the suction side of the circulating pump. Because the expansion tank compensates for pressure changes within the closed loop, it eliminates the negative pressure, or vacuum, caused by the pump. It is not absolutely necessary to follow this suggestion, but your system will have optimum circulation if you do. The expansion tank should always hang below the pipe it is attached to. Heat will accelerate the deterioration of the bladder inside the tank, and by hanging it below the pipe, the tank will stay cooler.

A pressure gauge is located in the closed loop to monitor the pressure within the loop. This gauge should hang down from the pipe and be visible from the charging valves. By hanging it below the pipe, you prevent air from getting trapped in the fitting or gauge. Unfortunately, most pressure gauges are designed to be above the pipe, so when you install one hanging below the pipe, the scale on the gauge will be upside down. Don't panic, you will get used to it.

A pressure-relief valve is also fitted to the closed loop. Note that this is a pressure-only relief valve. The relief valve should be set to open at 90–100 pounds per square inch or be set to the lowest rated component in the system. Be sure to install a drainpipe on the relief valve and terminate the drain near the floor. Some designs show the pressure-relief valve located on the roof. It is better to have it inside the building, with a drop pipe ending near the floor. Put a pail under the relief drop pipe. If the system owner or maintenance person sees fluid in the pail, that will be a clue to call the service person because something went wrong.

The solar loop circulating pump has no static head pressure to overcome in a pressurized system because all the pipes are filled with fluid at all times. The only head pressure the pump has to overcome is friction head. Therefore, low-head circulating pumps can be used to circulate the solar fluid through the solar loop. A PV-powered pump can be used in this system because the pump can start slowly and circulation will start immediately. In fact, a PV-powered system is best for several reasons. Most importantly, a PV pump runs whenever the sun is shining, so there is always circulation in the solar loop when the collectors could be hot. This extends the life of the solar fluid by eliminating stagnation. The only time temperatures in the 180°F range are encountered in these systems with PV power would be in water heating systems during no-load times such as vacations during the summer. Dangerous conditions are possible only in hot climates under this scenario, and a drainback system is probably preferable where this would occur. A PV-powered pump naturally runs at variable speeds: when the sun's resource is low, the pump runs more slowly. The thermal collector's output is also lower under this condition, so pump speed is perfectly matched to thermal collector output. On the other side of the coin, under full sun conditions, the pump runs faster and matches the temperature of the thermal collector by increasing the flow. Care should be taken with PV-powered systems when corrugated stainless steel pipe is used in the solar loop, because this type of pipe creates significant friction head to the flow, and pump sizing must be carefully calculated.

When installing a PV-powered system, it is important to carefully consider the location of the PV panel. The PV panel must be located in a completely shade free location. Any shading on the panel, however slight, will disrupt the output from the panel and result in erratic pump operation. If the system is located in a climate that experiences prolonged snowy conditions, it is advisable to mount the PV panel in a location where it can be seasonally adjusted. In this scenario, mount the PV panel so it can be tilted to 90 degrees at the start of the snowy season and adjusted back to your latitude angle the rest of the year. The vertical position will eliminate snow buildup on the PV panel and will ensure consistent operation all winter.

A differential temperature controller can also control pressurized systems.

There are differential temperature controllers for both DC- and AC-powered systems. Using a controller on a DC-powered system usually has the advantage of adding system monitoring and digital readouts so the system can be monitored better. In an AC-powered system an AC-powered controller is connected to a low-head 120-volt AC pump, which is used in the solar loop. A complete description of how these controllers operate is included in Chapter 4. If a differential temperature controller is used in pressurized systems, we suggest disabling its high-limit function to assure that the pump will circulate at all times when the collector is hot. If the power goes out during a sunny period and a controller is used to regulate the pump, the pump will stop and stagnation can occur. If this occurs for a long time, deterioration of the solar fluid may occur, so make a note to yourself to check the condition of the solar fluid more often.

Venting air

When filling the system with solar fluid, it is essential to get all the air out of the system. As mentioned above, the solar fluid, once installed, will be in the pressurized

Figure 5.3: *AC-powered pressurized antifreeze system*

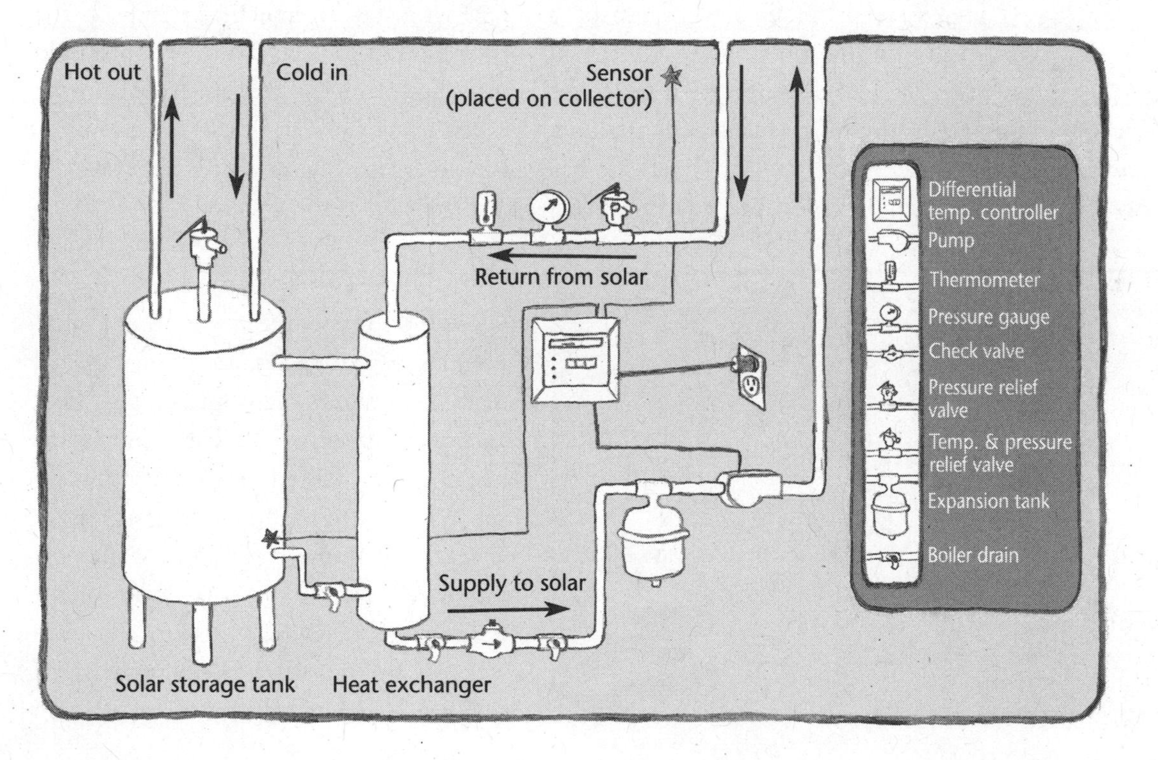

system for at least ten years under normal circumstances.

Air can cause circulation problems, so doing a good job of getting all the air out of the solar loop at the start will result in many years of trouble-free service. (The system-charging process is detailed in Chapter 9.) Many installers over the years have specified placing an automatic air vent at the highest point in the closed loop. If you choose to do this, it is important to follow these recommendations. First, always use a special "solar" valve that has a high-temperature float and a cast body. Install a short riser at the high point of your system on the collectors. Install a ball valve on that riser, and install the air vent on the end of the riser. Air will accumulate in the riser because it is the high point in the loop. You can open the ball valve, open the air vent to expel any accumulated air, and then close the ball valve. Instead of an air vent, you can also install a Schrader valve or a coin vent above the ball valve to accomplish the same task.

Figure 5.4: *Schrader valve air release*

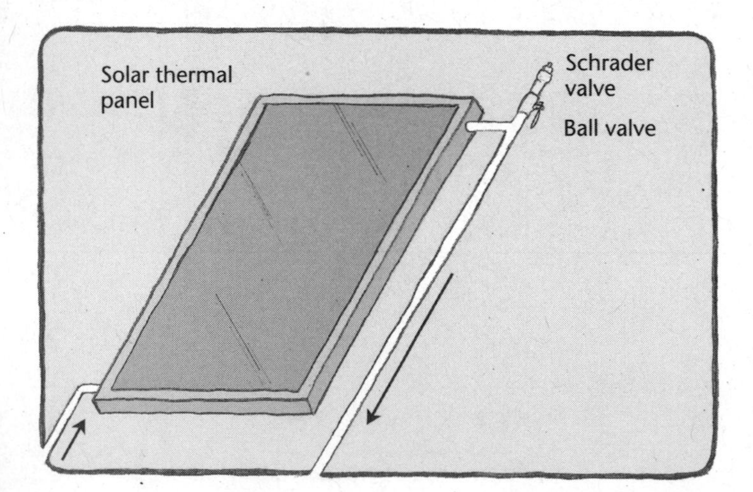

Some installers who are used to installing traditional hydronic heating systems will want to install an automatic fill valve on a solar closed-loop system. They do this out of habit because most hydronic heating systems require this valve to keep their system pressurized. Never install an automatic fill valve on a closed-loop system. It will destroy the system by diluting the solar fluid. Diluted fluid can lead to decreased freeze protection. It is acceptable to install an injection pump on the solar loop that injects pre-mixed solar fluid into the solar loop. Extreme care must be taken when calibrating this injection system to ensure that excessive pressures are not built up in the system.

Heat Exchangers

Closed-loop systems can use either a tank-integrated heat exchanger or an external heat exchanger. Tank-integrated heat exchanger systems are the simplest to install and are suited for situations where space is very limited. These systems are always single-pumped systems, as the only pump required is the solar-loop pump. External heat exchanger systems are more efficient and can also be single-pumped if the heat exchanger is a thermosiphon type. A plate-type heat exchanger or a coiled tube-in-shell heat exchanger can also be used. In these cases, the system must be double pumped. Double-pumped systems can be either PV-powered or use a controller and two 120-volt AC pumps. Remember to calculate the correct wire size when running long, low-voltage wires between the PV panel and the DC pumps. A complete discussion of heat exchanger types is included in Chapter 4.

Drainback Systems

Drainback systems are one of the three most popular types of solar energy systems installed worldwide. They are an excellent choice for all climates except those that experience severe or extended cold conditions, or where a significant amount of snow is expected annually. In warm climates that experience few, if any, freezing episodes per year, they are a good choice.

A drainback system works very well in warm climates because when the storage tank gets heated to its maximum desired temperature (its high limit), the system turns off and all the fluid drains out of the system, which prevents the solar fluid from degrading because of overheating. This is especially the case when systems experience idle periods such as vacations during the summer months, in space heating systems or in hot climates. It is also good in large commercial water heating systems where the heating load is not consistent. Because water — the best heat transfer fluid — is most often the solar fluid in drainback systems, these systems perform very well and at high efficiency when designed and installed properly. In severe climates, such as those experienced in many parts of Wisconsin, a weak antifreeze solution is suggested. Installers in this climate use a 30 percent glycol mix as insurance against catastrophic failures.

A drainback system is classified as an indirect solar water heater because the domestic water is heated by a solar fluid and heat exchanger, and not heated in the collector. It is classified as closed-loop because the solar fluid remains within a single circuit at all times. It is classified as an active system because it always uses a pump or pumps to circulate fluids through the system.

The major limitation of drainback systems is their inability to prevent freezeups in climates with extended cold or snowy conditions. Extreme care must be taken in these climates when installing pipe runs to ensure quick and complete drainage. Pump sizing is also critical.

Another limitation is that the collectors have to be located above the drainback tank. This eliminates the option of ground-mounted arrays. It should also be noted that the high-head pumps required to operate these systems use more electricity than the small pumps used on closed-loop antifreeze systems. Also be aware that pump selection is more critical in a drainback system and there are height restrictions on how high a high-head pump can lift water. Unless the circulating pump has a motor of more than one-half horsepower, the distance between the water level in the drainback tank and the top of the collectors must be less than 28 feet.

A drainback system uses flat plate or evacuated tube collectors to heat a solar fluid. Two insulated pipes connect the collectors to a specialized tank called a drainback tank. A high-head pump is installed on the pipe that feeds the collector array. A differential temperature controller turns the circulating pump on and off. When the pump is on, solar fluid is pumped from the drainback tank to the collectors, where it is heated. The solar fluid then drops back down to the drainback tank

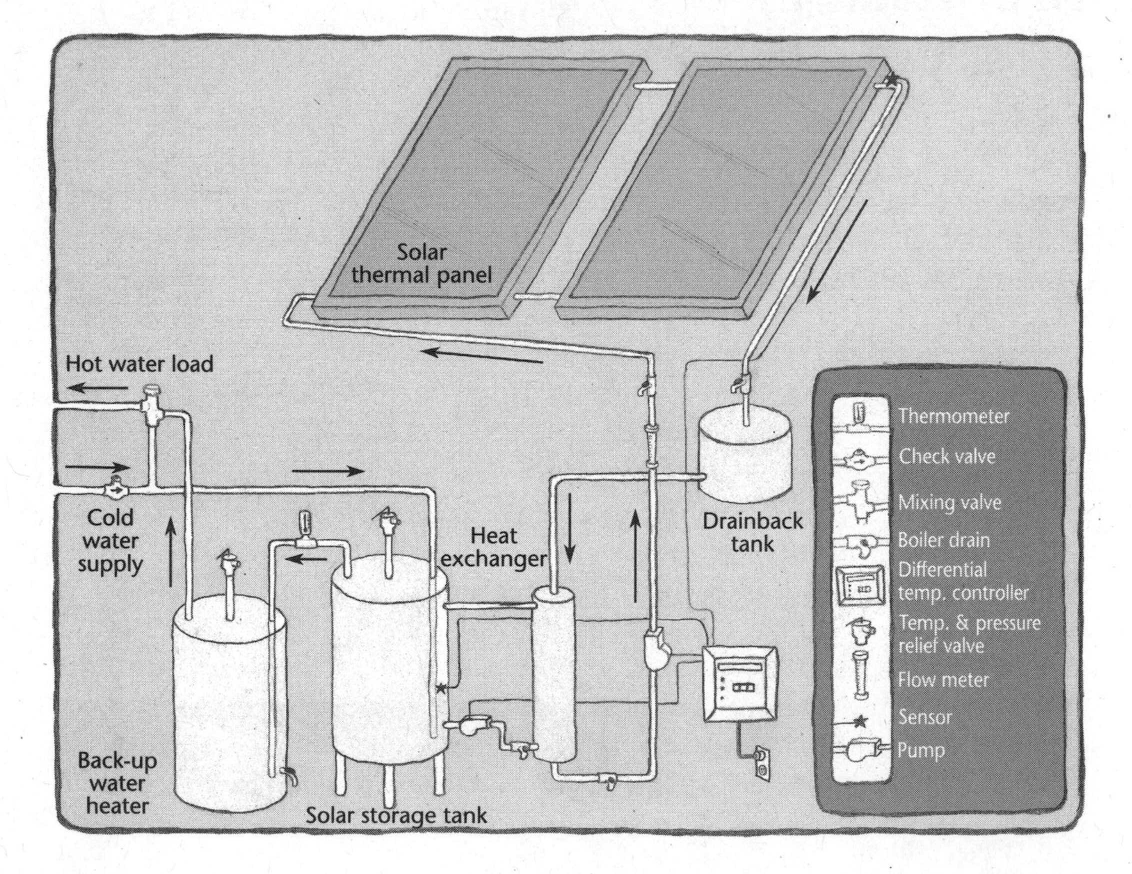

Figure 5.5: *Drainback system*

completing the circuit. When the system turns off, the pump stops, and all the solar fluid in the collectors and in the piping drains back into the drainback tank. Drainback tanks are relatively small, so they do not store much heat. When the system is operating, various methods, described below, are used to transfer the heat from the drainback tank to a solar storage tank.

Drainback systems are used for residential water heating systems and residential space heating systems. They are also used for commercial water heating systems and space heating systems.

Sizing a drainback system is exactly the same as sizing a pressurized antifreeze system. System sizing is covered in Chapter 7.

To make sure that drainback systems function properly, a few rules must be adhered to:

- The system must be installed to facilitate fast, complete drainage when it turns off.
- The collectors must be mounted so that they drain toward the inlet of the array; this would be the bottom manifold, where the solar fluid enters the collectors.
- They should be mounted at a ¼-inch-per-foot slope toward the feed inlet.
- Collectors should never be mounted so that the riser tubes of the absorber plate are horizontal, as they will sag, over time, and prevent proper drainage. If sagging occurs, water will be trapped in the pipe and cause it to burst during freezing conditions.
- Collectors with serpentine riser tubes cannot be used in drainback systems.
- Some evacuated tube collectors cannot be used in drainback systems. Refer to manufacturer specifications to ensure that the collectors you have chosen will work in a drainback system.
- All piping must be sloped at a minimum of ¼-inch-per-foot slope toward the drainback tank, and all horizontal pipe runs must be supported at least every four feet to prevent any sagging, which would inhibit proper drainage.
- Use a pair of 45-degree elbows instead of a 90-degree elbow whenever practical to facilitate faster drainage.
- The minimum pipe size for drainback systems is ¾" hard copper.
- Corrugated stainless steel pipe can not be used in drainback systems, only rigid pipe.

Drainback tanks should always be located within the conditioned space. They can be located at the highest point within the conditioned space to reduce head pressure. This is an important consideration when the system is installed in a two-story house and the storage tank is in the basement. Drainback tanks should always be unvented. All drainback tanks should be fitted with a sight glass to monitor fluid levels within the tank. The only exception to this would be where a low pressure drop flow meter is used in the feed line as the sight glass. To size a drainback tank, calculate all the liquid that would fill the collectors and all the piping above the drainback tank, and add four gallons. These tanks should be well insulated to prevent heat loss. A pressure-relief valve should be mounted on the drainback tank. Note that many drainback tanks have a maximum pressure rating of 50 pounds, so set the pressure-relief valve to that pressure.

Figure 5.6: *Flow meter*

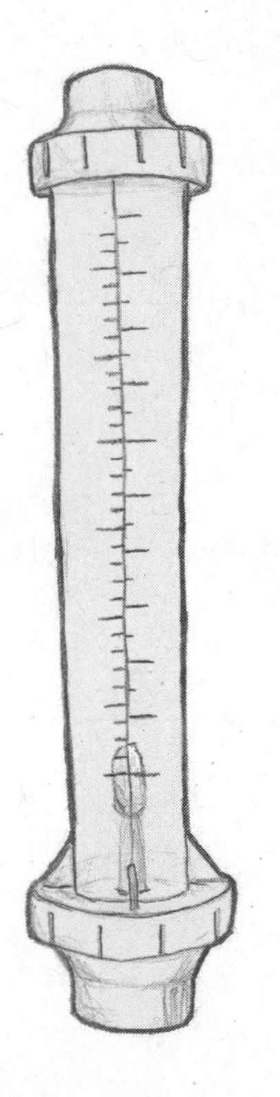

The solar-loop pump must be a high-head pump of sufficient size to pump the water from the drainback tank to the top of the collector array. When calculating the head, measure from the bottom of the drainback tank to the highest point of the collector array, and add four feet. Your pump must be able to exceed that head plus the friction head of the piping. Once all the piping is full of solar fluid, the pump does not have to work very hard because gravity pulling the fluid back down the return line helps pull fluid up the feed line. It will take a 120-volt AC pump to do this job, so the system must be powered by 120-volt AC and use a differential temperature controller to turn the pump on and off. These systems cannot be PV powered because when the system turns on, the pump must start with full force to overcome the head pressure, and PV-powered pumps don't work that way. The pump should be located below the bottom of the drainback tank and be located in a vertical pipe, pumping up to the collectors. Sizing the pump is critical because you need a robust pump to overcome the total head when the system is filling but must also not pump the solar fluid too fast once the head is overcome. There are special pumps called high-head, low-flow pumps that work best in these systems. Some systems have used either a variable-speed pump or a booster pump placed on a timer to minimize energy usage once the static head has been overcome.

A drainback system includes a storage tank to store the solar heated water for later use. There are various methods of getting the heat from the drainback tank to the storage tank, all of which use a liquid-to-liquid heat exchanger. One method circulates the hot solar fluid through an in-tank heat exchanger (including the wrap-around type). In this design only one pump circulates the hot solar fluid throughout the system. This type of system is called a

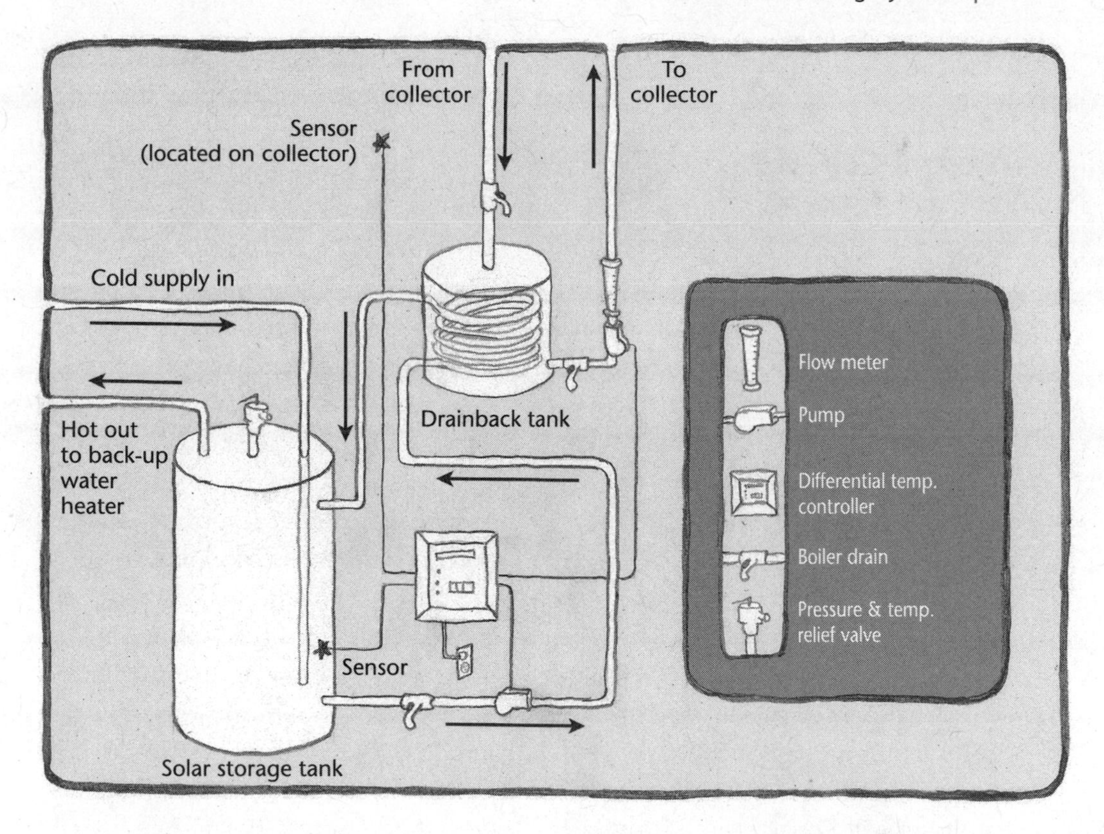

Figure 5.7: *Drainback system with tank heat exchanger*

single-pumped system because there is only one pump.

Another method of getting heat from the drainback tank to the storage tank is to have a heat exchanger below the minimum fluid level inside the drainback tank. There are commercially available drainback tanks with a submersed heat exchanger. A second pump is required with this system to circulate water from the storage tank and through the heat exchanger and back to the tank. This style is called a double-pumped system because it uses two pumps. Both pumps are controlled by the differential temperature controller and turn on and off at the same time. The heat exchanger pump should be very small because the flow through that circuit should be slow and because there is very little total head to overcome. Too large a pump will destratify the tank.

A third method uses an external heat exchanger mounted near the storage tank. This configuration operates similarly to the example above in that it would also need a second pump for the domestic hot water side. This is the design shown in Figure 5.5.

Integral Collector Storage Systems

An integral collector storage (ICS) solar water heating system is considered passive because it requires no pumps of any kind for operation. It is considered a direct system because the domestic water actually enters the collector. The ICS unit is typically plumbed in series between the cold water supply and the conventional water heater. Whenever a hot water tap in the dwelling is opened, cold water from the supply enters the ICS collector and forces the solar-heated water stored there into the conventional or backup water heater. If the water from the ICS collector is hotter than the setting on the backup heater, that heater will not activate. If the water from the ICS collector is warmed but is below the temperature setting of the backup heater, the backup heater will then have to add only enough heat to bring the water up to the preset temperature. If no solar heating has taken place and all the water in the ICS collector is cold, the backup heater will have to deliver the whole load.

Because water is plumbed through the ICS collector, this type of system is suitable only for climates or seasons where there is no chance of freezing conditions. ICS collectors are very popular along the extreme southern parts of the US and in tropical climates. ICS collectors have also been successfully used for summertime use only, typically in conjunction with vacation homes and recreational facilities such as parks and campgrounds where they are used only during the summer months and are drained the rest of the year.

Because of their simplicity, ICS collectors are among the least expensive solar water heating systems available. There are no pumps, extra storage tanks, controllers or other components needed. A complete system consists of the ICS collector and piping. Because of its simplicity and low initial cost, people who live in climates where freezing conditions do occur are sometimes tempted to use this system. They often think that if freezing conditions are forecast, they will simply drain the water out of the collector and piping and everything will be all right. In theory, this assumption is correct. In reality, Murphy's Law — if anything can go wrong, it will — often comes into play. Actually, all ICS collectors can withstand mild freezing conditions. The mass of the water in the collector will keep it from freezing for quite some time (depending on the temperature). It is the piping going to and from the collector that is vulnerable. A well-insulated ¾" copper pipe will freeze in less than five hours at 29°F.

Most ICS solar water heaters do not suffer because of hard water. Only in areas with extremely hard water is there a potential problem. Most ICS solar water heaters have large tanks and waterways, decreasing potential mineral buildup problems. Also, the tanks do not tend to get excessively hot, decreasing mineral deposit buildup.

If you live in a climate that does not experience freezing conditions and you do not have excessively hard water, this is the system for you. If you need hot water only during the summer months, for a campground or summer cottage, for example, the ICS solar water heater is a viable option for you as well.

Because ICS systems actually store water within the collectors, these collectors weigh the most of any type of solar collector. Water weighs more than 8

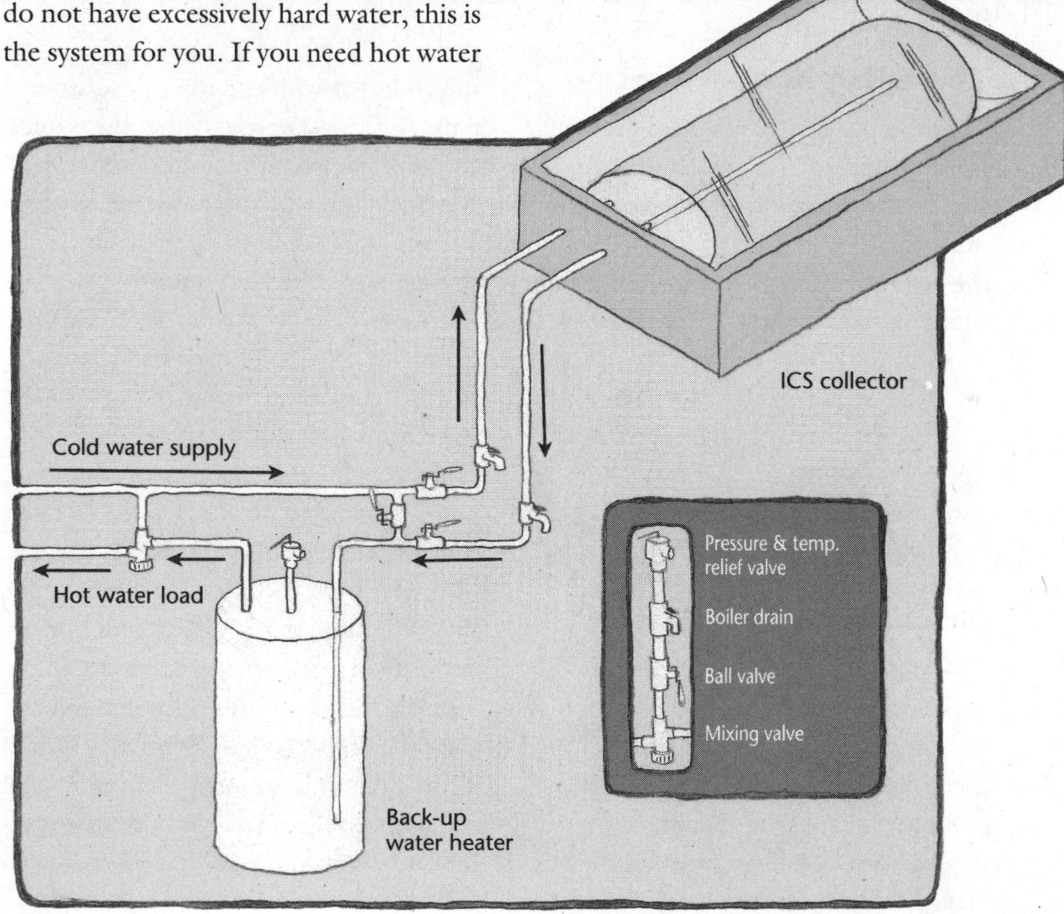

Figure 5.8: *ICS system*

pounds per gallon, so a 40-gallon ICS solar water heater can easily weigh more than 500 pounds. You may need to make provisions if this type of collector will be roof mounted. Most buildings can support an ICS solar water heater with no modifications, but be sure to check out your situation before going ahead. It is no fun having a collector full of water — and perhaps a few installers — come tumbling through your roof.

Thermosiphon Systems

Thermosiphon systems all use the principle that hot water naturally rises. In this system the storage tank is located above the collectors so that as fluid is heated in the collectors, it rises to the storage tank. These systems are constructed similarly to ICS systems, except that you can construct them out of individual components instead of a single collector unit. Two systems use the thermosiphoning principle. Both are considered passive systems because they don't require pumps.

Direct Thermosiphon Systems

A direct thermosiphon system typically uses a flat plate collector and heats domestic water directly by circulating it through the collector. It also uses a storage tank to hold the solar-heated water. Usually the bottom of the storage tank is connected to the bottom of the collector with a pipe, and the top of the storage tank is connected to the top of the collector with a pipe. Because the tank is located above the collector and is filled with water, the collector is always full of water. When the sun comes out, it heats the water in the collector. At this point, the water in the collector is warmer than the water in the storage tank. Because the water is warmer than the water in the storage tank, it will rise up to the top of the storage tank through the pipe that connects the top of the collector with the top of the storage tank. As this water rises within the system, cooler water from the bottom of the tank is drawn down to the collector through the pipe that connects the bottom of the storage tank with the bottom of the collector, setting up a circulation without the use of pumps. The most efficient way to heat this water would be to use flat plate or evacuated tube collectors. However, because the system is direct, you will be sending hard water through the collector. In most cases this is a bad idea. Flat plate collectors have small riser tubes that are easily clogged with mineral deposits. If such a system were to be installed, a water softener located prior to the collector would be essential. Though a direct thermosiphon system is one of the simplest systems, if your location permits it, you would be better to choose an ICS system.

Indirect Thermosiphon Systems

Some indirect thermosiphoning systems have been manufactured over the years. Like the thermosiphon system described above, the indirect system also has the storage tank mounted above the collector. Instead of circulating the domestic water through the collector, a heat transfer fluid is used, which eliminates problems caused by hard water. When the solar fluid is warmed in the collector, it rises naturally to the level of the storage tank, where a heat exchanger transfers the heat from the solar fluid to the domestic water. A chamber where the solar fluid is circulated often surrounds the storage tank. By circulating the hot solar fluid through this chamber, the water tank it surrounds is heated. The system can also be installed using a sidearm tube-in-shell heat exchanger like that seen in Figure 5.9. The collector in this type of system is protected from freezing because an antifreeze solution is used as the solar fluid, but the piping serving the storage tank could still freeze.

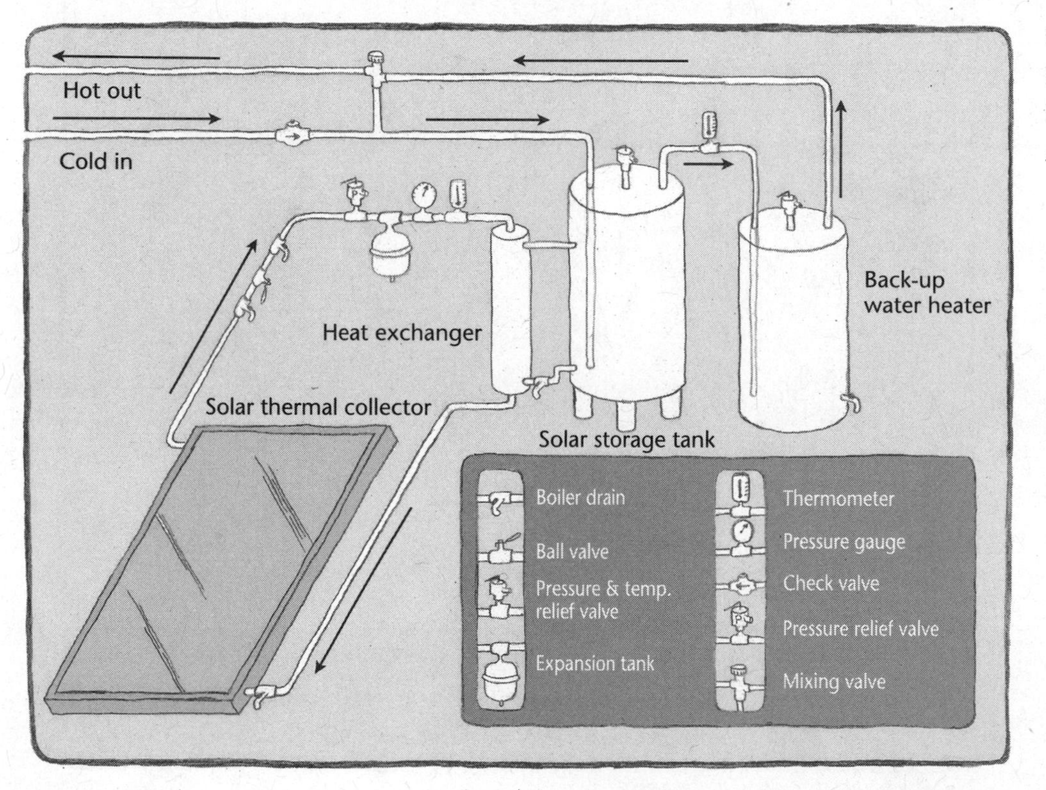

Figure 5.9: *Indirect thermosiphoning system*

I have seen this system successfully installed. If the storage tank is located in a heated area, you eliminate freezing problems. However, configuring the system in this manner is not suitable for most locations. Where it has worked, the collectors were mounted on the overhang of a porch, and the storage tank was located on the third story of the home. Typically, the utility room is located in the basement and not on one of the upper levels. Most homeowners aren't willing to sacrifice the amount of area that it takes for the rest of the system. For this reason, indirect thermosiphoning systems are usually not the best option.

Open-Loop Systems

Flooded Open-Loop Systems

A flooded collector open-loop system has water in the collectors and piping at all times. This system is composed of a storage tank, a collector (or collectors), a circulating pump, a controller and piping. The

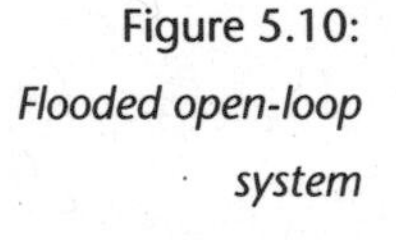

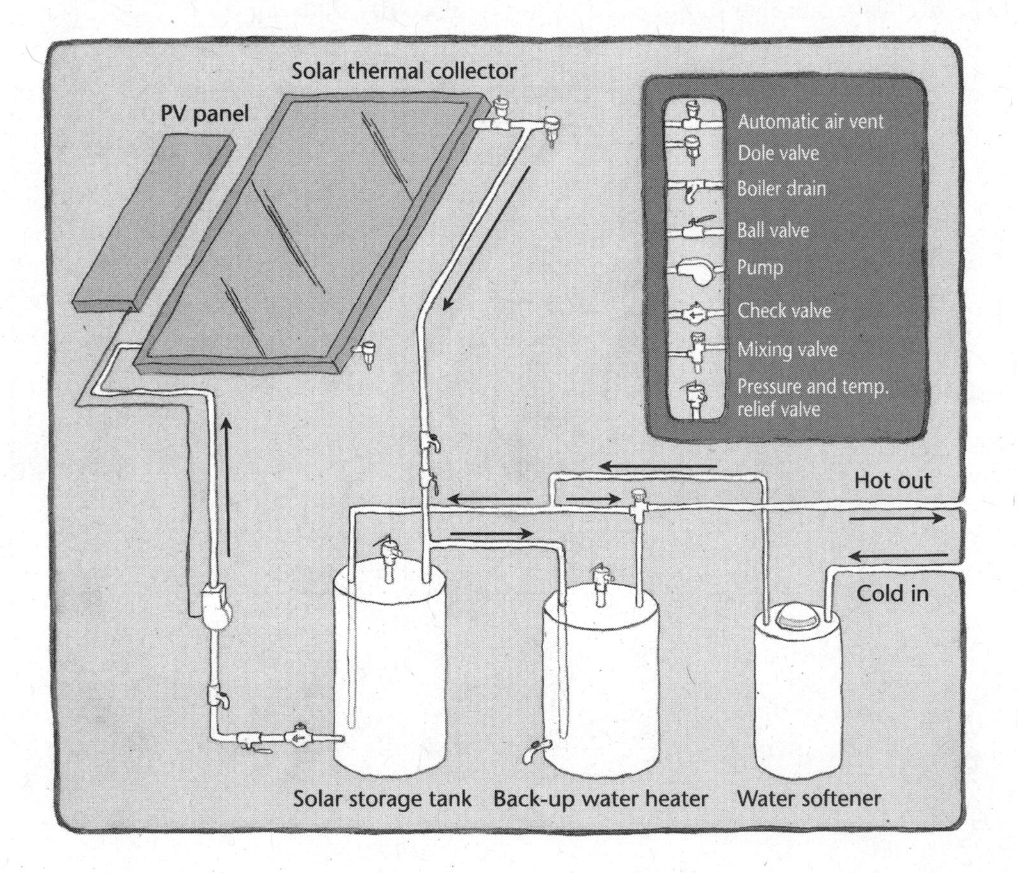

Figure 5.10: *Flooded open-loop system*

collector is plumbed directly to the storage tank. When solar energy is available for heating, the pump turns on. Water is circulated from the bottom of the storage tank to the collector, where it is heated. The water, after being heated directly in the collector, is then circulated back to the storage tank. At the end of the day, the pump simply turns off and the circulation is stopped. These systems can use a microprocessor-based controller with sensors to control the operation of the pump. A simpler system uses a photovoltaic (PV) solar collector to convert solar energy directly into electricity to operate the circulating pump. A PV-powered system eliminates the need for a controller because the only time the pump will run is when solar energy is available.

These systems typically use flat plate collectors but can also use evacuated tube or concentrating collectors. Although they are more complicated than passive systems, they are still less expensive and simpler than most other active systems. Because most of these systems use flat plate collectors, there are more options for collector placement than ICS systems.

Because these systems are flooded, they are vulnerable to both freezing conditions and mineral buildup in the collectors. They have no freeze protection and should not be used in locations that ever experience freezing conditions. In addition, hard water is a concern, and a water softener located before the collector is essential.

Recirculation Systems

Recirculation systems are like open-loop flooded systems, but they have the added feature of being able to recirculate warm water from the solar storage tank throughout the system if freezing conditions are experienced. Recirculation systems must be microprocessor controlled. If freezing conditions are experienced, the controller turns the pump on and water is circulated throughout the system, reducing the likelihood of freezing. Many differential temperature controllers have a recirculation function. This system really works only during mild and occasional freezing conditions, so where extended freezing conditions are experienced, it is not appropriate. Hard water issues must be considered as with any flooded system as mentioned above.

Another negative of a recirculation system is that heat stored in the storage tank is lost during the recirculation process, seriously decreasing overall efficiency.

Draindown Systems

All draindown solar water heating systems are direct systems because the domestic water is heated directly in the collector. A draindown system is made up of a collector (or collectors), a storage tank and piping to connect the collector and storage tank to the home piping; it may also use an electronic control, various valves and a pump or pumps. Typically, when solar energy is available, a specialized valve, such as a Sunspool, opens and allows household water to fill the system. Then a pump, controlled by a microprocessor with sensors, turns on and circulates cold water from a storage tank to the collectors. The water is heated as it passes through the solar collector and then returns to the storage tank.

Some systems use a photovoltaic collector to power the circulating pump, thus eliminating the need for the controller and an external energy input. When the sun sets or is blocked by clouds, the pump shuts off and the water that was in the collector(s) is allowed to drain down out of the collector and piping. This small amount of water is piped to a floor drain and is lost.

Draindown systems are very efficient because the water is heated directly, as in an ICS system. Because draindown systems

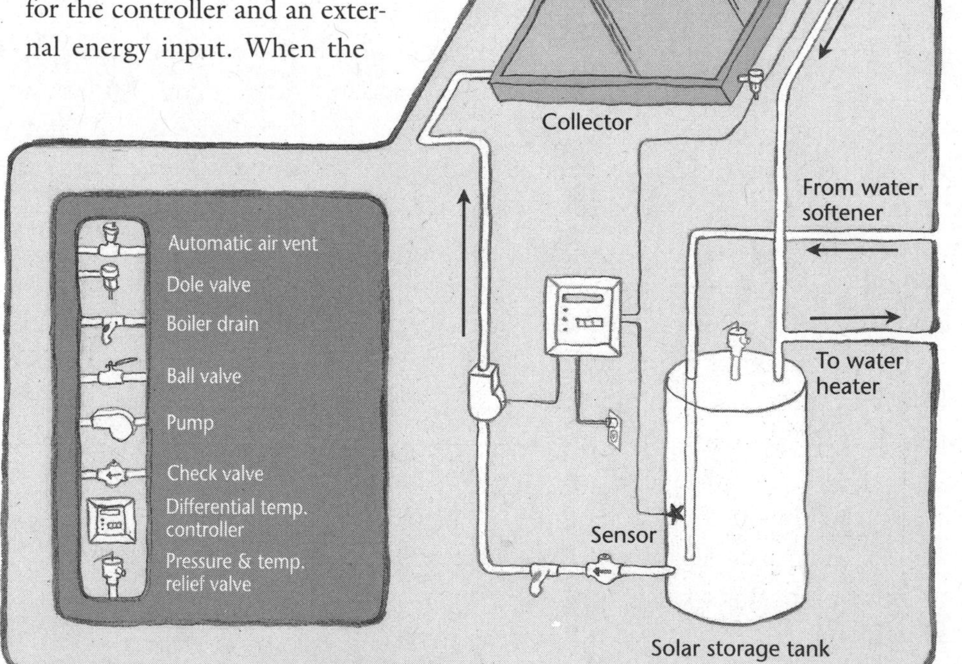

Figure 5.11: *Recirculation system*

use flat plate collectors, these systems are even more efficient than ICS systems. But, like any system that heats the water directly, there is the problem of mineral buildup in the collectors. As mentioned above, this is a serious consideration, and if there is any mineral content to your water, this type of system is not recommended. The specialized valves necessary for this type of system have been prone to failure. Because of this type of system's tendency to freeze, it is not recommended for climates where freezing conditions may occur more than once or twice a year.

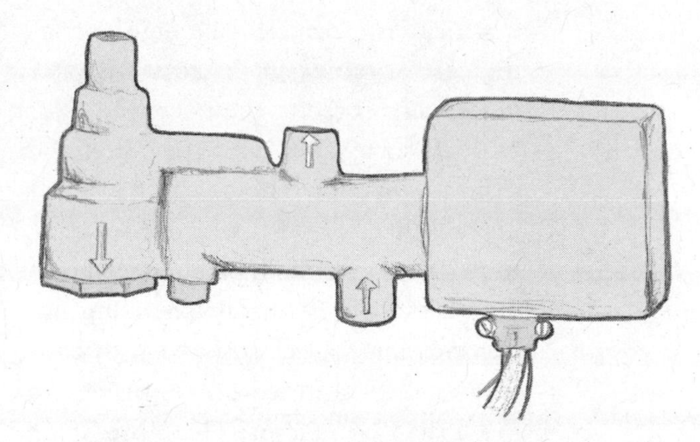

Figure 5.12: *Draindown valve*

Refrigerant Solar Water Heaters

Heat pumps work on the principle of latent heat, by which a fluid absorbs energy (heat) when the liquid vaporizes. When the fluid condenses, heat is released. In a solar energy system, a fluid called Freon absorbs heat in the collector as it evaporates, and then it releases that heat in the solar storage tank.

Because heat pumps can work at very low temperatures, they can extract heat from low-quality heat sources. For instance, during cloudy weather, when most liquid solar collectors will perform poorly, a refrigerant system can concentrate the low-grade heat for water heating. Some manufacturers of refrigerant solar water heaters have claimed they don't even need the sun to work: their systems absorb heat from the surrounding air. When operated like this, they are very similar to the standard home heating heat pumps that are popular today in moderate climates.

Most heat pumps power a compressor with an electric motor that can be several horsepower. Heat pumps are rated by their coefficient of performance (COP), a ratio of the energy that was moved divided by the energy it took to accomplish the task. Most heat pumps can deliver about three times the energy required to operate the system. Put another way, a heat pump requires about one kilowatt-hour of electricity (3,400 Btu) to move 10,000 to 13,000 Btu. This equals a COP of 2.5 to 3. An electric heater at 100 percent efficiency has a COP of 1 because it delivers one Btu of energy for each Btu it uses.

So this sounds pretty good, doesn't it? Might something be wrong with this picture? Let's look a little deeper. First of all, electricity is expensive. The high operating cost of refrigerant-based solar water heaters compared to the operating costs of other solar water heaters has been the main reason this technology hasn't gained in popularity. Second, this method of using the COP to rate the system misses a crucial calculation. Electrical generating and transmission losses associated with conventional fossil-fueled power plants are more than 66 percent. Therefore, if we factor in the losses of generating and delivering electrical energy, the COP of a heat pump system is really only 1, and is similar in efficiency to a gas water heater. Third, as mentioned above, most heat-pump systems use Freon as the refrigerant. When released into the atmosphere, many types of Freon reduce the ozone layer that protects the earth from harmful solar radiation. Heat pumps lose efficiency as the temperature drops. Attempting to operate heat pumps in very cold weather can result in damaged equipment.

Heat-pump solar heating systems have been promoted by electrical utilities. They use the initial COP rating to classify these systems as being energy efficient, and compared to resistance heating, they are. But as we have seen, they are no more efficient than burning gas. A few companies continue to sell heat-pump-based solar energy systems. In reality, these systems should be classified as solar-assisted heat pumps because they are really just old-fashioned heat pumps that have been given a slight boost by solar input.

In some recent developments, the buildings to be heated were built to the very highest standards for insulation values and very low overall heat loss. Ground-coupled heat pumps were installed to heat these homes and, because of their small size, they could be powered by a large PV system. It is an interesting concept that is quite costly.

Solar Heated Pools

Two kinds of pools can be solar heated: seasonal swimming pools and indoor, year-round swimming pools. We will also briefly discuss solar hot tub and spa heating systems.

Seasonal Pools

More square feet of solar swimming pool heating collectors are installed per year than all other solar thermal collectors combined. In fact, seven pool collectors are sold in the US for every flat plate or evacuated tube collector. These are popular solar heating systems because they are cost effective. These systems use unglazed collectors made of a special polypropylene plastic. The piping is traditionally PVC. They are usually draindown systems and therefore are direct systems. These systems typically consist of the solar collectors, the piping, a controller and a diverter valve. The pool filter pump usually circulates the fluid through the collectors, so no additional pump is required.

Seasonal solar swimming pool heating systems can eliminate the need for fossil-fueled heating systems. A solar pool heating system can raise the temperature of your pool up to 15 degrees above ambient temperature. If you use a pool cover, which we highly recommend, you can raise the temperature by another ten degrees. Most solar pool heating systems will pay for themselves in two or three years in the savings by not using fossil fuels. This is a 30 percent to 50 percent return on investment!

The solar pool heating controller has two sensors, one to measure the temperature of the pool water and one to measure whether solar energy is available. The controller usually has a dial that sets the desired pool temperature. Most controllers work in conjunction with the filter timer,

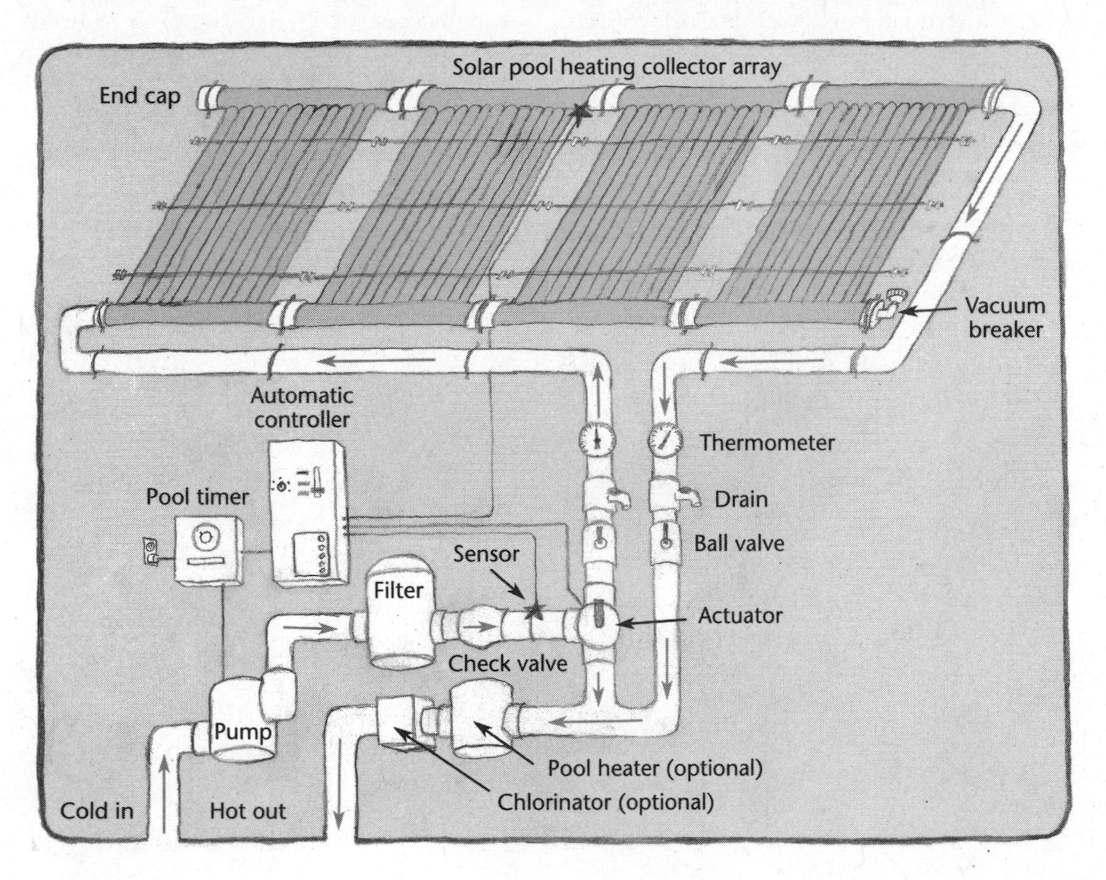

Figure 5.13: *Seasonal pool system*

which is set to operate during the day. When solar energy is available and the pool needs heat, the controller turns on a motorized diverter valve, also called an actuator, so the pool water is directed to the solar collectors.

As the pool water flows through the collectors it gains heat, usually about one degree Fahrenheit with each pass. After exiting the collectors, the water returns to the pool. The diverter valve is located near the exit of the pool filter, so the water is filtered before it enters the collector loop. Pool filter pumps are usually large enough to easily pump the water through the collector loop. However, in some cases a booster pump may be required.

The diverter valve is a three-way valve. When solar energy isn't available, or when the pool reaches its desired temperature, the controller returns the diverter valve to its normal position. In this position, water flows straight through the valve and goes from the filter directly to the pool. In the solar heating position, the valve diverts the filtered water to the collector array. When the solar heating system turns off, all the water that was in the collector array and the solar-loop piping drains into the pool. In order to facilitate complete drainage, a vacuum breaker is typically installed at the bottom end of a collector array on the opposite end to where the water enters.

As we mentioned in Chapter 3, plastic collectors are preferable for seasonal pool-heating applications because most pool water is treated with chlorine, which is not compatible with copper. The collectors can be unglazed because high temperatures are never required and because they are used only when the ambient outdoor temperature is above 50°F. Each collector manufacturer has its own method of fastening the collectors to a roof or mounting structure. These mounting systems typically consist of special strapping to hold the collectors tight to a roof or flat surface. Please refer to the instructions that come with your system, as they all vary slightly. As mentioned in the collector descriptions, we like the web-and-tube collector configuration best. This allows the collectors to expand and contract without kinking or binding. This configuration is also easy to repair. These collectors will last 15 to 20 years, and most manufacturers offer long warranty periods.

Follow manufacturer recommendations for collector array sizing, layout options and pipe sizing. For commercial or large residential pools, be sure to contact the manufacturer for its recommendations. Collector array sizing is based on climate, pool surface area, desired water temperature, amount of shading on the pool, orientation of collectors and use of a pool cover. In general, the size of your collector array will equal between 50 percent and 100 percent of the pool's surface area.

Indoor Solar Swimming Pool Heating Systems

Indoor swimming pools can be heated year round with solar energy. In most cases, if the solar heating system is designed to provide about 50 percent of the heat during the winter months, it will provide about 100 percent during the summer months without overheating the pool. These systems are cost effective because they are used to their capacity every sunny day of the year. Indoor pools have exactly the same heat load all year. The reason the systems provide less heat during the winter is that the solar resource is less then. It is typical to size the collector array to heat the pool on the best solar day of the year. This sizing method ensures the pool will never overheat.

These systems are typically pressurized or drainback systems. They are installed just like solar water heating systems and solar space heating systems with the exception of the heat exchanger. Special heat exchangers are required because pool water is corrosive to copper. Large swimming pools have surge tanks to hold excess water from the pool when many people jump in the pool at the same time. These surge tanks sometimes contain the filters. Pex tubing can be coiled inside the surge tank to act as the heat exchanger. For pools that do not have a surge tank, the heat exchanger is located in the filter piping directly after the filter. With pressurized systems you cannot just turn off the collector loop like you can with seasonal pool heating systems when the pool reaches your comfort temperature. The collectors will be on every sunny day, so sizing is more critical to avoid overheating the pool.

Pressurized systems can be PV powered. Only one pump is required—the solar-loop pump. For pools with a surge tank, that is the only control needed. For pools without a surge tank, the pump filtering system and pump must be set to run during daytime hours. Because the year-round solar resource varies widely from one climate to another, we always use a computer-modeling program such as RETScreen to help with array sizing.

The payback on indoor year-round solar heating systems is a little longer than for seasonal pools because the equipment is more expensive and the systems must be slightly undersized to prevent overheating during the summer months. Mounting the collectors at an angle to maximize the winter sun's resource can reduce the probability of overheating. Sizing is calculated by considering the pool's surface area, the rated collector output, the desired pool temperature and the use of a pool cover.

Solar Heated Hot Tubs and Spas

Hot tubs and spas can be good candidates for solar heating. This is especially true if you already have, or anticipate getting, a solar water heating system or a solar space heating system. The kind of hot tub or spa you have will dictate the kind of solar heating system that is appropriate for your application. The biggest cost in heating these items comes from filling them and warming the water initially. Maintaining the temperature in a well-insulated tub does not take as much energy. It makes a lot of sense to insulate your tub well.

Bob has a wooden hot tub outside his home that holds about 600 gallons of water. The tub is part of the shunt load for his solar space heating system. The tub serves two purposes: it is a place to put the excess heat generated by the solar heating system during the non-heating part of the year, and it also is a luxury item that the whole family thoroughly enjoys. For half the year the tub is heated 100 percent by solar power, and it is tempered all year round. The tub has wooden benches set around the inside near the bottom. Bob placed 100 feet of 1-inch Pex tubing under the benches and diverted the solar fluid through the tubing. This heat exchanger does a nice job of heating the tub and keeps the fluid cooled. Many of the systems we have designed over the years utilize a tub like this as a shunt load.

There are two approaches to using solar energy to heat hot tubs and spas: a dedicated solar heating system that serves the tub or combined tub/spa heating with solar water heating. Only wood-stave (Ben calls them pickle-barrel) hot tubs lend themselves to dedicated solar heating systems. One 4' × 8' collector works well to heat and maintain a wood-stave hot tub. A heat exchanger can be added to the tub as described above. A simple closed-loop system is set up. PV power and a DC pump is best. Plastic hot tubs and spas do not lend themselves to dedicated solar heating systems because it is impossible to get a heat exchanger into the water, as the inside is molded plastic. And because these spas are kept hot all the time, when the sun is out and the tub is at optimal temperature, there is little heat load per hour and there is no place to dump heat without overheating the unit. Therefore, the only practical method is to heat a reservoir during the day with solar energy and then take heat from the reservoir when needed.

Under normal circumstances, the tub or spa will be associated with a building that has a domestic water-heating load. I suggest incorporating the tub/spa heating system with a solar water heater. As in swimming pools, spa and tub water is often treated with chemicals that are not compatible with copper.

All piping that comes in contact with spa water should be schedule-40 PVC solid core pipe. Any heat exchanger that comes into contact with tub or spa water should be marine-grade stainless steel or cuprous nickel. Most plastic tubs or spas are well insulated, so once these are hot, they do not have a large heating load. Usually one 4' × 8' collector is all that is needed to maintain the temperature. By adding one panel to your solar water heating array and adding about 30 gallons of storage capacity to your solar water heater, you will be most of the way to completing a system. You will also need a heat exchanger, as mentioned above, a small 120-volt AC circulating pump and two aquastats to get the heat from the solar storage tank to the tub or spa. Install the heat exchanger in the plumbing line directly after the tub or spa filter so the filtered water flows through the heat exchanger before it returns to the tub or spa. Run insulated copper piping from the top of the storage tank to the hot inlet of the heat exchanger and from the cold outlet of the heat exchanger to the bottom of the storage tank. Install the circulating pump on the cold line, pumping from the exchanger to the tank. Wire the system by tapping into the 120-volt AC line coming out of the timer on the tub or spa that runs the filter pump. Run that line to the first aquastat, then to the second aquastat and then to the circulating pump. Set the first aquastat to turn on at temperature drop and set the temperature to five degrees above the setting on the tub or spa heater. Set the differential at three degrees. Put the sensor for this aquastat so that it measures the water temperature. Set the second aquastat to turn on at temperature rise and set the temperature at five degrees above the tub or spa temperature and set the differential at three degrees. Place the sensor of the second aquastat to measure the temperature of the tank about one third of the way down from the top of the tank. Whenever the

Figure 5.14: *Pickle-barrel hot tub*

tub or spa filter is engaged, the second aquastat will recognize if there is heat in the storage tank to give to the tub/spa, and the first aquastat will make sure the tub or spa will not get too hot from the solar heating system.

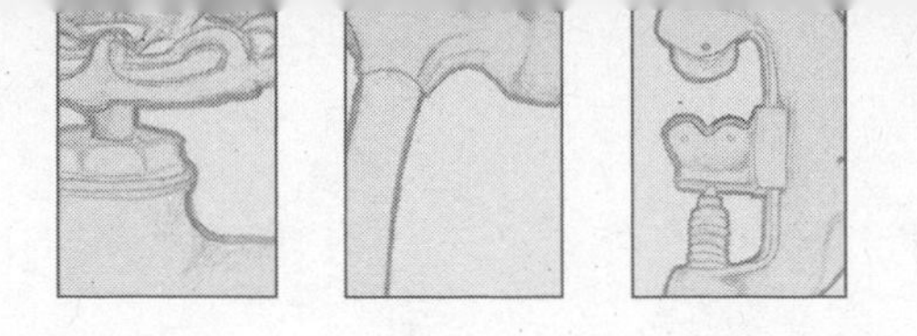

6

SOLAR SPACE HEATING SYSTEMS

IN MANY PARTS OF THE WORLD, the average household spends a significant portion of its annual energy bill on space heating. In our quest to use less fossil fuel, we can look to the sun to help reduce our dependence on those finite resources.

This chapter deals with active solar space heating systems. An active system is one that uses a solar collector to capture the sun's energy. Passive solar heating utilizes the building itself as the solar collector through the use of south facing windows and thermal mass within the building. Although it is a great idea and should be a requirement for all new construction, we will not deal with passive solar heating in this book.

This chapter will outline the available types of active solar heating systems and will detail the systems that have proven to work the best and are the most practical. The type of system you might consider will depend on the kind of building you have, how that building is used and the climate it is located in.

Unfortunately, most buildings were built to minimum energy standards, and that trend continues with new construction. Whether you are building new or retrofitting an existing building, you will get the best return on your investment by making energy conservation the first step in reducing your energy bills. It makes economic sense to do whatever is possible first to increase your building's energy performance and reduce its energy demand, and then size a solar heating system to address that reduced heating load. In most areas you can get a home energy audit where a professional energy efficiency expert can let you know how to reduce the heating load.

It is rare that a solar heating system can provide 100 percent of a building's annual space heating requirements. There are usually many cloudy days during the heating season, and it is impractical to have a solar collector array or a heat-storage mass large enough to hold enough heat to do the job. In a high-performance building, however, it is possible to provide a majority of the annual heating with solar energy, and in an average building a significant amount is possible. How far can we go with solar heating systems? For all systems except high-mass radiant systems, something just around 50 percent of the yearly total space heating requirement is a practical goal. High-mass systems can provide more of the annual heating load for a building, even in very cold climates.

Active solar space heating systems can be classified as those with heat storage and those without storage. When we spoke about solar water heaters, we addressed the fact that the sun shines directly into our collector for only three to eight hours a day. If we want solar-heated water to use after the sun goes down, we must incorporate some way of storing the heat collected during the day for later use. We also acknowledged that the sun does not shine every day, so we want to store that solar heat for as long as is practical. Some climates, like ours here in central Wisconsin, can experience a whole week of cloudy weather during the early winter months. During the whole month of December, we may only have five or six sunny days.

Another factor to keep in mind when sizing solar heating systems is that the solar energy system's output is variable. Likewise, the heating demand of the building will vary greatly over the whole heating season. If we size the solar heating system to heat the building 100 percent on the coldest day of the winter, then on sunny days that are not as cold we will have excess heat and could overheat something. If we size the solar heating system to heat the building on an average sunny winter day, during the worst weather we will not quite get all our heat from the solar heating system, but over the whole length of the heating season we could get a significant amount.

Solar space heating systems can also be effective without storage. We call this type of system a dump system because heat is delivered directly from the solar collector to the heated space; it is dumped there for immediate use. In most instances, the interior of the building and its contents actually act as a heat-storage medium, so even in this type of system there is some minimum amount of storage.

The two most popular types of collectors for space heating are air-type collectors and liquid-type collectors. These types of collectors have been described in Chapter 3. Both have been used with success for solar space heating. Each has its advantages and disadvantages and best applications.

Liquid-Type Solar Heating Systems

Liquid-type solar heating systems gained popularity during the early 1980s and continue to be a popular option today. Several new innovations since the '80s make this an even more attractive and practical choice today. We can break this type of system into two basic types: systems with storage and systems without storage.

As mentioned in Chapter 5, a solar water heating system is in most cases the most cost-effective renewable energy system. Solar space heating systems are usually the next most cost effective. If a building has a domestic hot-water load, your first choice in solar energy should be a solar water heating system because that system is used year round. If you want to use solar energy to heat your building and you will have a solar water heater, the most practical choice is a combination solar water heating and space heating system. Whenever a solar thermal system is a combination system, the collectors are to be tilted to maximize the winter sun.

Liquid-Type Systems Without Storage

We call systems without storage dump systems. In this case, a liquid is heated in the solar collector and is plumbed into the building. The hot liquid is then directed to some type of radiant panel, radiant floor or a liquid-to-air heat exchanger. Familiar types of liquid-to-air heat exchangers include baseboard heating units, old-fashioned radiators and fan convectors. Details about heat delivery methods are given later in this chapter. If this is a combination system, the excess heat generated by the collectors can be dumped into the building for space heating.

Systems without storage add heat to a building only when the sun is shining and the system is turned on or in the heating mode. These systems should be sized to heat the building on an average winter day, where the output of the collector array is matched to the heat load of the building. If the collector array is too large, the building could suffer from overheating, or the system itself could overheat. The solar collectors should be oriented to maximize the low winter sun angle so their output will be the highest during the heating season and decrease in the spring and fall. These systems do not have to work at all during the non-heating season, so the angle of the collectors can be very steep or vertical. This orientation will minimize the potential to overheat the solar fluid during the non-heating season and will help match the collector output to the heating load of the building.

Drainback Solar Heating Systems Without Storage

The solar loop for a drainback solar heating system without storage is exactly like the solar loop used in a drainback solar water heating system. A common type of drainback tank to use with this system is the kind that does not have a heat exchanger inside the tank. Whenever heat is available, the primary pump circulates water from the drainback tank and through the heat delivery system, dumping heat into the building.

The advantage of this type of dump system is that once the building gets up to its upper limit, the system can simply turn off. Also, during the non-heating season, the system simply stays off and there is no concern about overheating the solar fluid.

For a combination drainback solar water heating and space heating system, the solar loop and the storage loop are configured like a plain drainback solar water heater except that the collector array is increased in size to accommodate the heating load. Of course, all the piping and the drainback tank size would be increased proportionately to account for the increased array size. This configuration will heat the storage tank rather quickly because the collector array is oversized for the storage tank (which is sized for the domestic load only). An aquastat is added to the storage tank and set at 130°F. When the storage tank gets above 130° a pump comes on and either delivers hot water from the storage tank to the heat delivery equipment or delivers heat to a heat exchanger, where the heat is transferred to another liquid that goes to the heat delivery equipment. Note that some plumbing codes allow domestic hot water to be used in a heating system as long as a) the piping is approved for potable water, b) there is no chance that the water will be contaminated by another liquid and c) the water is circulated through the radiators periodically. In other words, you could not send domestic hot water into a hydronic heating system that was heated by another source. Some codes never allow domestic hot water to be used for space heating, so in those areas, the second heat exchanger is required. We do not recommend using domestic water in a heating loop.

A thermostat is placed in the line between the aquastat and the heat circulating pump and is set to open on temperature rise. This thermometer is used to control overheating and should be set at the maximum allowable temperature. When the building reaches that temperature, the thermostat will turn off the pump.

Pressurized Solar Heating Systems Without Storage

Pressurized systems without storage follow the same description given above for the drainback systems with the following exceptions. If the antifreeze system is not a combination system, the solar fluid can be sent directly to the heat delivery equipment. Pressurized systems cannot just be turned off like the drainback system can, so care must be taken to keep the antifreeze from overheating. Adjusting the tilt angle of the collectors will help, but oversizing the collector array can be a problem, especially in warm climates where the heating demand is small during the winter. We also suggest not using the high-limit thermostat in the building; just let the system run. If it gets too hot, open a window. A diversion load should also be supplied to shunt excess heat during the summer to keep the solar fluid from overheating. Common diversion loads include buried pipes or radiators placed outside the building; they are detailed later in this chapter.

Liquid-Type Systems With Storage

Liquid-type solar heating systems with storage are simply expanded solar water heating systems. There are two kinds of heat-storage mediums that are used with liquid-type systems: water and sand.

Heat-Storage Configurations

In this section we will describe two kinds of water-storage configurations: large single tanks and multiple smaller tanks plumbed together to create a volume of liquid equal to one large tank. The large single tank is almost always an open or atmospheric tank, which means that the lid can come off and is not sealed to be a pressure vessel. Both large storage tanks and multiple smaller tank systems can be configured for pressurized systems and drainback systems.

Large Single-Storage-Tank Configurations

The best solar storage tank is seamless and jointless. With no seams or joints, there is less chance of a leak as the tank ages. The best tanks will also be constructed of a material that can withstand consistent 180°F temperatures. The tank will also have to be insulated, and it must have a tight-fitting lid that does not drip. Another important consideration, especially in retrofits, is that you must be able to get the tank into the building.

Tanks have been constructed of almost every conceivable material that can hold water. Many site-built tanks were wooden frames lined with plastic or rubber roofing materials. None of these tanks lasted more than 10 or 15 years — and some as few as two years — without developing leaks. Steel oil drums of 250-gallon capacity or larger have been used as solar storage tanks. These tanks were never made to contain water, so they developed leaks over time as well.

Our favorite solar storage tank is made of high-temperature-rated fiberglass. A number of these tanks are on the market and they are reasonably priced (at the time of printing) at about $2 to $3 per gallon of capacity. These tanks come in a variety of sizes. The 400-gallon size is 40 to 48 inches in diameter and 4 to 6 feet high. The tank must come in pieces in order to get it through doorways. The tank is then assembled in place. Some tanks are bolted together using a rubber gasket to seal the joints; others are cold-welded using special epoxy cement. The epoxy-sealed joints are the best.

Being from the Dairy State, we have seen a number of used stainless steel bulk milk tanks used. These have actually worked very well because the stainless steel does not corrode and the pieces are welded together, eliminating potentially leaky seams. You can also have your own stainless steel tank constructed. This can be costly, but it will last a long time. We have done this when there was a finite amount of space in which we could fit the solar storage tank. You can get more gallons per square foot with a square tank than a round one. Rectangular ones also fit through doorways more easily. If you have a tank like this constructed, be sure to have plenty of reinforcement and cross-bracing.

Recently we have seen some successful installations using pre-cast concrete septic tanks. These are typically quite large and are meant to hold water. Concrete septic tanks must be buried and must be waterproofed on the inside with a cement/latex waterproofing plaster. These tanks cannot be freestanding as they will crack when filled with water.

The most trouble-free tanks are those with no fittings, which are always potential leak sites, especially on round tanks. This means that all the piping to and from the tank must come out of the top of the tank. However, all tanks require a drain, so there will have to be at least one fitting. If there has to be a fitting on the tank, have it installed at the factory. Field-installed fittings are unreliable, especially on round tanks. Because most configurations also require at least one heat exchanger inside the tank, you must have access to the inside of the tank through a lid of some sort, and the bigger the better. The lid should have a flange on the inside to prevent condensation from dripping over the side.

Most large tanks do not come insulated, so you will have to install insulation. Set the tank on at least two inches of extruded foam. Wrap the tank with reflective insulation and fiberglass bats, with the foil on the outside. We like to then wrap the insulation with some type of covering to make it more durable and also look nice. Reflective silver bubble wrap products work well for this purpose. Use foil tape to make a nice-looking job. You can also have the tank sprayed with foam insulation.

Tank heat exchangers are often submersed coils of Type K soft copper or Pex tubing suspended within the tank. Note that copper coils are four to six times more efficient than an equivalent length of Pex tubing when used as a heat exchanger. Copper-finned tubing can also be used as a heat exchanger. Sizing considerations for these heat exchangers are covered in Chapter 7.

Typically, multiple heat exchangers are placed inside the tank, depending on the application. Most common is to place near the bottom one coil that is connected directly to the solar loop of either a drainback or pressurized system and is used to dump heat into the tank. Near the top of the tank is typically one coil that is used to draw heat off for domestic hot water and another coil that is used to pull heat for space heating purposes. In this case, the tank acts as a heat transfer station, and the water that is actually in the tank never leaves.

These heat exchangers can be set on a rack inside the tank or they can hang from the top. You can make a rack out of ½-inch rigid copper tube and a few fittings to hold the heat exchanger off the bottom and also hold any exchangers that are near the top. Be sure to put caps on the bottom of the legs so they won't wear a hole in the bottom if they move around through expansion and contraction.

Using multiple coils is the easiest and most trouble-free method of extracting heat from your tank, but you can also use the hot water directly from your storage tank in space heating applications. To do

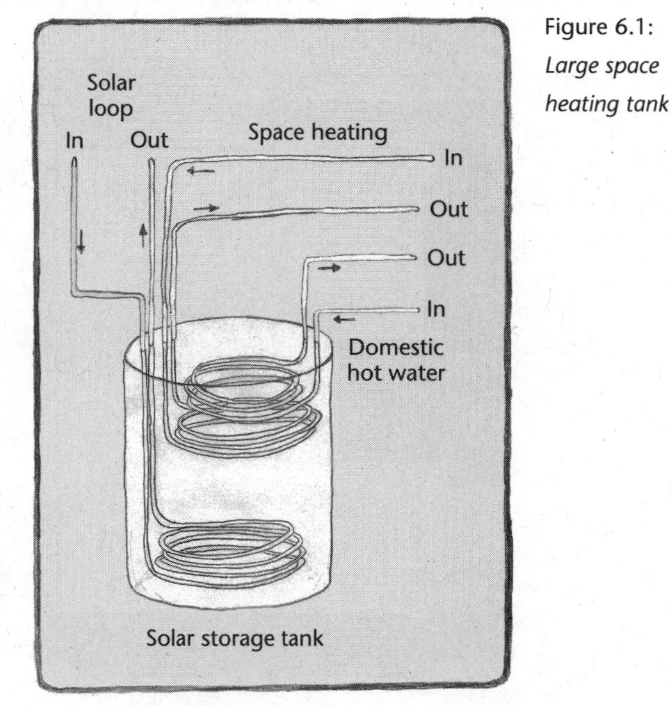

Figure 6.1: *Large space heating tank*

this you need to install a bulkhead fitting on the side of the storage tank a minimum of 16 inches below the water level in the tank. A pipe attached to the bulkhead fitting on the outside of the tank turns down and runs along the side of the tank to near the bottom, makes a 180-degree turn, and then runs in an upward direction for a short way. The heating-loop circulating pump is then installed on this pipe, pumping away from the tank. The pump must be located below the water level in the tank to keep it primed. If you need a bulkhead fitting on a fiberglass tank, get a solid brass one, or better yet, have one installed at the factory. Do not over tighten plastic or fiberglass fittings that are sometimes found as bulkhead fittings. We will discuss the ways to operate the heat extraction heat exchanger in the part of this chapter that deals with heat delivery methods.

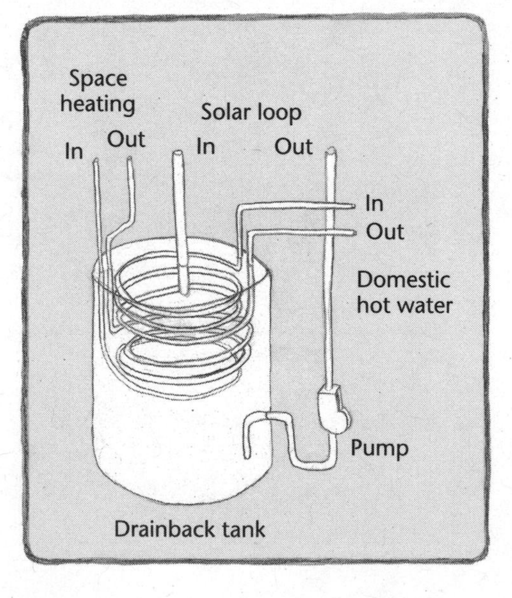

Figure 6.2: *Drainback space heating tank*

Although rarely used today, some drainback systems use the fluid inside the large tank as the solar fluid. The large water tank stores the heat and is the drainback tank. A bulkhead fitting is installed on the side of the tank at least 16 inches below the tank water level, and a pipe configuration exactly like the one just described above is installed. In this case the solar-loop pump is a high-head pump capable of overcoming the static head of the system. The solar fluid leaves the tank and travels down and then up into the high-head pump. It is then pumped up through the collectors, where it is heated. After exiting the top of the collector array, the heated water then drains back to the large storage tank. The static head is calculated from the water level in the storage tank to the top of the array plus four feet. In cold climates the whole large tank must have antifreeze in it.

Multiple-Storage-Tank Configurations

As stated above, multiple tanks can be used to store heat instead of one large storage tank. Multiple-tank heat-storage configurations can be used for both drainback and pressurized systems. In either case, the solar loop is used as the primary heat source.

Whenever multiple tanks are used, they should be plumbed in parallel so they all heat up at the same rate. Likewise, when we take heat out of multiple tanks we should take it out evenly by having the tanks plumbed in parallel. There is a special technique to plumb multiple tanks in parallel; it is called reverse-return piping. It is sometimes also called first-in, last-out plumbing. It is the same technique used when plumbing solar collectors together. In this piping system, imagine a series of tanks set next to each other, with each tank having a hot and cold port on the top. The cold feed line has a pipe that runs to each tank, and this pipe connects first to tank #1 (first in), then to #2 and then to each subsequent tank. For example, if there were three tanks, the first tank in line would be the first tank to be attached

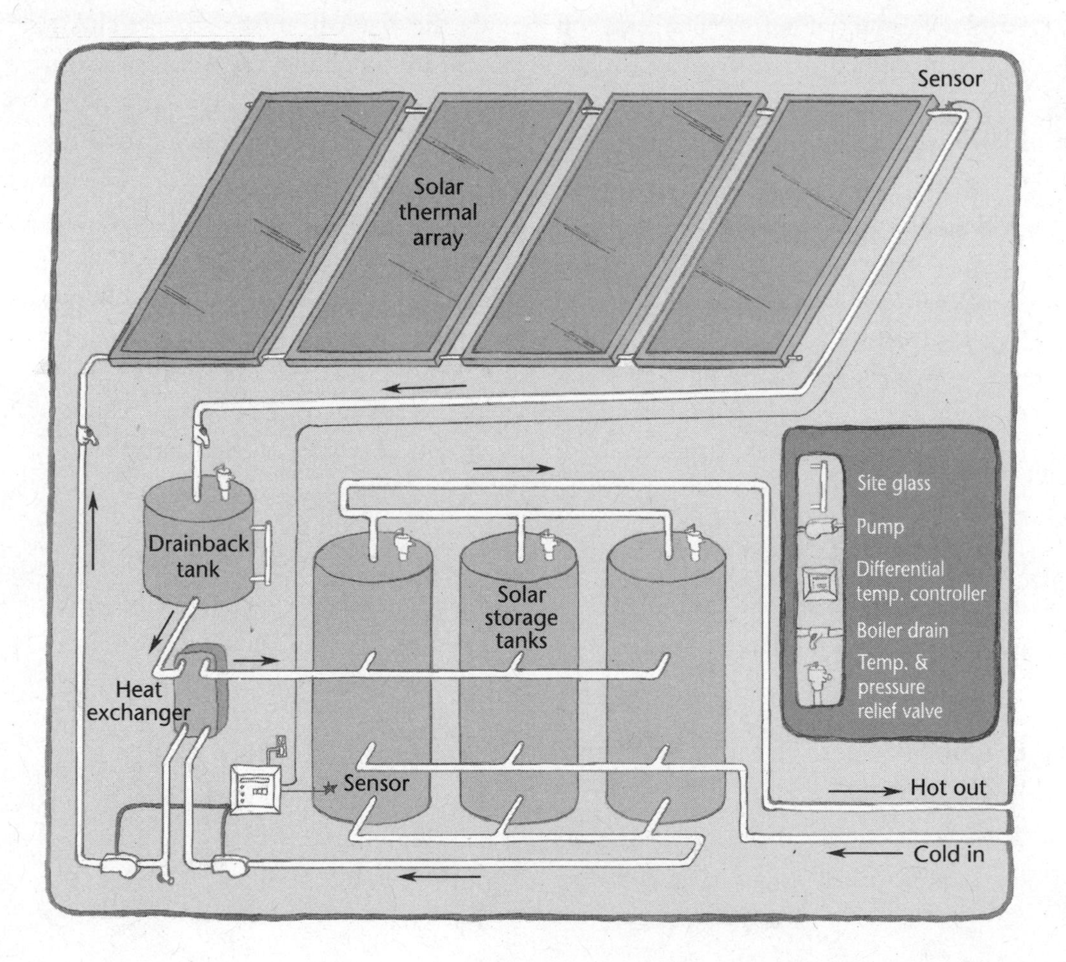

Figure 6.3: *Drainback system with multiple storage tanks*

to the cold line and would be the first in, or closest, to the water supply. The third tank would be the last in line to get cold water (last in). The hot out line also has a pipe that runs to each tank, but the tank closest to the outlet would be the third tank, and the farthest tank from the outlet would be tank #1. This configuration is used in Figure 6.3.

One way to heat the tanks is to use tanks with integral heat exchangers. You can then plumb the solar loop to pass through all the heat exchangers in the individual tanks. Plumb the tank heat exchangers in parallel, using reverse-return configuration. You can also use standard storage tanks and a single heat exchanger to extract heat from the solar fluid and put it into the tanks. In this case, there will be two parallel sets of pipes installed on the tanks, one pair to put heat into the tanks and one pair to take heat out. Both pairs should be plumbed reverse-return. Using multiple tanks will cost at least twice as much as the complete cost of a single fiberglass tank, and the fiberglass tank will last longer than individual steel tanks.

Drainback Solar Heating Systems With Storage

The solar loop for a drainback solar heating system with storage is very similar to a drainback solar water heating system solar loop. There are several options: one uses a single tank that acts as a drainback tank and heat-storage tank; one uses a conventional drainback tank in conjunction with a large storage tank with heat exchanger; and one uses a conventional drainback tank and several conventional steel heat-storage tanks with internal or wraparound heat exchangers.

In systems with a large single storage tank, the water from the storage tank is pumped through the collectors when solar energy is available. The high-head pump is located as low as possible next to the tank and below the water level inside the tank. When the system turns off, the water in the collectors and piping drops back into the large tank. The system uses a differential temperature controller to activate the pump. Whenever solar energy is available, the large tank gets heated. There is a high-limit setting on the controller, so when the tank reaches its desired temperature, the system turns off and the water drains from the system back to the storage tank. A heat exchanger is submerged in the tank to extract heat for the heating system. A combination system would have a second heat exchanger in the storage tank to heat the domestic hot water.

This configuration's main limitation is the likelihood of its freezing if installed in a climate with prolonged or severe cold conditions or large amounts of snow annually, so this type of system is not recommended in such a climate unless the tank has an antifreeze mixture in it. Another limitation is the distance from the collectors to the storage tank. It needs a circulation pump large enough to overcome the static head from the level of the storage tank to the top of the collector array. This could be a serious issue if the building is more than two stories tall. As with all drainback systems, the collectors must be located above the storage tank to facilitate proper drainage. This eliminates the possibility of remote-mounting the collector array.

The big advantage of these systems is their ability to simply turn off and drain when the storage tank reaches its upper limit. This is especially advantageous in warm climates or where the load is not consistent over the course of a week.

For systems that use multiple storage tanks, a conventional drainback tank is used, and solar fluid from the drainback tank is circulated through the tank-integrated heat exchangers on each individual tank. Under this scenario, the water held in the storage tanks is pumped directly to the heating system. If this is a combination system, either the water in the storage tank is used for both purposes or an additional heat exchanger is used for the domestic water heating. If a heat exchanger is used, there will have to be an additional circulating pump and control to circulate hot water from the storage tank through the heat exchanger.

This design does not suffer from the height issues that limit the large-tank systems because the drainback tank can be mounted a considerable distance above the storage tanks. However, it does need to have the collectors above the drainback tank and is limited by the same climatic conditions as all other drainback systems. Drainback solar heating systems can use sand as the heat-storage medium. We will discuss this kind of system in detail later in this chapter. These systems have a traditional solar loop, and like the drainback systems with large storage tanks, the solar loop extends into the thermal mass sand bed to deliver the heat from the collectors directly to the storage medium.

Pressurized Solar Heating Systems With Storage

Pressurized solar heating systems with storage offer three storage options: systems that use one large storage tank, systems that use multiple storage tanks and systems that use sand as the storage medium.

The solar loop for all three storage methods is exactly the same and follows the traditional pressurized design that was described in the solar water heating section.

The traditional method of storing heat for these systems uses a single large storage tank. The solar loop delivers heat to the tank through a heat exchanger located in the storage tank. Heat is extracted from the storage tank through other heat exchangers inside the tank. Multiple heat exchangers can be placed inside the tank, so combination systems are simple to install. The solar loop for these systems can be PV powered or can use differential temperature controllers. This is the preferred system for retrofit applications and is compatible with many existing heat delivery systems. A diversion load is typically required to shed excess heat that may be collected during the non-heating season.

This system's big advantages include versatility of collector placement, absolute freeze protection for all climates and regulated temperature control inside the building. The disadvantage is the system's tendency to overheat when installed in warm climates, especially when the load is intermittent. This limitation can be overcome by having a diversion or shunt loop installed in the solar loop.

When multiple storage tanks are used, the water in the storage tanks is usually sent directly to the heat delivery system if possible, but additional heat exchangers can be used to extract the heat from the tanks. The easiest tanks to install and engineer are those with integrated heat exchangers. Combination systems require a separate storage tank for the domestic hot water. These systems can get quite complicated and expensive. The advantages and disadvantages for this style are the same as those listed for the pressurized system with a large single storage tank.

Pressurized solar heating systems can use sand as the heat-storage medium. We will discuss this kind of system in detail later in this chapter. These systems have a traditional solar loop, and like the pressurized systems with large storage tanks, the solar loop extends into the thermal mass sand bed to deliver the heat from the collectors directly to the storage medium. Combination systems are typical, especially in residential applications. Combination systems have an additional solar hot-water-storage tank, configured exactly like those in typical pressurized solar water heating systems. In this case, as the hot solar fluid is brought into the building, some of the hot solar fluid is sent to a liquid-to-liquid heat exchanger to heat the water in the storage tank and the rest is sent to the storage mass.

These systems differ from all other solar heating systems with storage because their amount of thermal mass is much larger. In fact, they typically have more than 50 times the weight of other systems. This very large thermal mass has the ability to store an appreciable amount of heat that tempers a building in a significant way. The advantages of this type of system are the same as those listed for the other pressurized systems, with the added advantage of not having a large storage tank within the building. No storage tank means no tank maintenance or possibilities of leaks. These systems can provide a majority of the annual heating load, even in extremely cold climates. Though overheating is always a possibility with any pressurized system, it is less likely with a high-mass system like this, if properly designed, installed and operated.

Diversion Loads

Almost all pressurized solar heating systems will have a shunt loop or diversion load. This diversion load is a place where we can send the excess solar heat when we do not want to use it in the building. We want to keep the solar fluid from getting extremely hot, thereby prolonging its life. We do this by sending the collected solar heat somewhere outside of the building.

After the hot solar fluid enters the building from the collectors and then passes through the heat exchanger for the domestic hot water, we divert it through the shunt before it goes back to the collectors. This effectively keeps the solar fluid from getting extremely hot. All solar fluids experience breakdown under overheat conditions, so if we keep the temperatures within tolerances, we will extend the life of the solar fluid and reduce costly maintenance.

Most shunt loops are a single long loop of Pex or copper tubing, which is buried outside the building. We suggest one linear foot of one-inch Pex tubing for each square foot of collector area for the shunt. Use one foot of one-inch copper for every two to three square feet of collector. If the soil is extremely dry at your location, the shunt won't work as well, so additional shunt length may be necessary. You can also use finned baseboard radiator tubing that is placed outside in a spot that is both shady and exposed to the wind. Placing it behind the collectors is sometimes the best option. You will want to use one linear foot finned tubing for every four square feet of collector area.

There are both manual and automatic diversion loop controls. Today it is easy to automate diversion loops because of the advancement in controllers. Some new controllers will automatically divert fluid when the storage gets to a preset high limit. It is also simple to configure an automatic diversion loop by placing a surface aquastat on the return pipe leading to the collectors that would be set below 180°F. When the high limit is met, the aquastat engages either an electronic shutoff valve, such as a zone valve, or a pump that diverts the hot solar fluid to the diversion load. Most zone valves operate on 24VAC, so you will

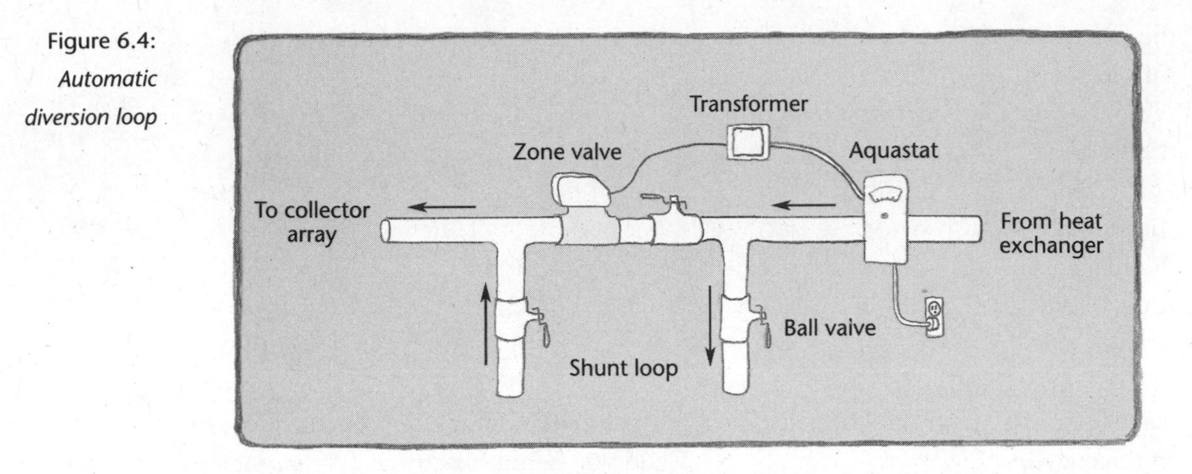

Figure 6.4: *Automatic diversion loop*

either need to run 24VAC power to the aquastat or use a step-down transformer to reduce the voltage from 120VAC. If a step-down transformer is necessary, we recommend placing it after the aquastat. Transformers will always draw power and are a parasitic load. The aquastat will prevent unnecessary electrical consumption.

In manual systems, diverting the solar fluid from the floor loops to the shunt loop is easy. There is a ball valve on the shunt loop and one on the solar loop. Simply close the solar-loop valve and open the shunt valve. This will manually divert all of the solar fluid through the shunt loop. Note: Never have both these valves closed at the same time because then the solar fluid will not circulate, overheating will occur and damage to the system can be expected.

Another option for a diversion load is to have a heat exchanger in an outside hot tub. This is the system used at Bob's home. He has placed a 100-foot length of 1-inch Pex tubing coiled around the inside of the hot tub near the bottom. When this diversion loop is engaged, the hot tub stays hot—it's great. We encourage you to be creative when designing the shunt. If you have any uses for some heat during the non-heating months, by all means try and use this excess solar heat. It's free!

Heat Delivery Methods

So far we have discussed methods of collecting and storing solar heat. How can this heat be used to heat your building? There are a number of options for taking the heat out of a water tank or using the heat directly from the solar collectors and delivering it to the building on demand. Which method you choose will depend on your existing heat delivery system, if there is one, and also might depend on the type of storage tank you have.

It is useful here to describe the different types of space heating systems. A wood-burning stove, for example utilizes both radiant heating and convection heating. If you stand near a hot wood stove, you feel warm on the part of your body that is facing the stove. That is radiant heat. The stove also heats the air around it and that warm air travels around the room. That is convection heat transfer. Convective heating systems heat air and the warm air fills the room to make everything in the room warm. A radiant heating system has a warm panel or floor surface that, when warmer than the surrounding air, gives off invisible electro-magnetic waves that travel out from the panel in a straight line until they hit a solid object. When they hit the solid object, it gets warm. Convective heating systems are less efficient than radiant because convective systems have to fill the room or building with heat. In order to get the floor warm, they have to overheat the ceiling. Because radiant heating systems heat things instead of air, the temperatures stay the same from floor to ceiling and that takes a lot less energy. It is also more comfortable because we live on the floor and it is typically nice and warm in radiant heating systems but typically cool/cold in convective heating systems.

Integrated and Non-Integrated Heat Delivery Methods

We mentioned earlier that a solar heating system typically is not the only heating system for a building. A solar heating system may be the primary heating system, but if we want the building to remain at a consistent temperature all winter, then we will need an additional heat source to fill in the blanks when solar is not available. As with domestic water heating systems, a backup heat source will be necessary. All types of heating systems are composed of a heat source and a heat-delivery mechanism. Ideally, we would like to be able to use the same heat delivery method for both the solar heat and the backup. This can save installation costs on either or both systems.

The first question is, What medium is used to get heat inside the space? Common media are air, water and electricity. The only heat delivery medium that is outright incompatible with solar is the electric resistance heater. You typically see it placed in the floor or along the baseboard. Because the medium is electricity instead of a fluid or air, there is no way in which the solar can bring heat to the space without an alternative method.

The next question is, What temperature does your system operate at? Heat delivery systems can operate at low temperatures or high temperatures, depending on the heat delivery equipment and the heat source. Low-temperature heat delivery systems require larger devices to deliver a set amount of heat, compared to high-temperature delivery systems, because the higher the temperature, the more concentrated the heat. For instance, radiant floors have a typical operating range of 90°F–130°F because they use the entire floor to deliver heat. This large surface area allows them to use a lower-grade heat source. Hydronic baseboard radiators have a typical operating range of 150°F–190°F. They require a much higher-grade heat source because just traveling around the baseboard of the room doesn't provide much surface area to deliver the heat.

Let's take a step back and look at the operating range of solar heating systems. Typically, you can expect to see temperatures ranging from 120°F–150°F at the end of the day. Occasionally we can experience higher temperatures, but only when there isn't a load on the system. Compared to the examples above, we should consider solar heating to be a relatively low-grade heat source. It should be noted that you can design your system to reach higher temperatures, but it requires significantly more collector area, decreases the system's overall efficiency and delivers a relatively small volume of high-temperature heat. Most systems are designed to provide large quantities of low-grade heat instead of small quantities of high-grade heat. Now, because solar energy is a relatively low-grade heat source, the best heat delivery system to use with solar is a low-temperature system.

Most heat delivery systems that use a high-grade heat source will be incompatible with solar. Consider the example of the hydronic baseboard radiators. Most systems are fed by a boiler that sends fluid out at 180°F. When it comes back, the temperature of the fluid has usually dropped to about 150°F. Remember that

Heating Systems Compatible with Solar	Heating System Incompatible with Solar
Domestic hot water	Electric resistance heating
Radiant floors, ceilings or walls	High-temperature boilers using radiators
Forced-air	Steam heating
Pre-heat of make-up air	
Industrial processes	
Crop drying	
Make-up water on commercial boiler systems	

our solar collectors are heating fluid only up to 150°F, so there is no place in the boiler loop in which we can tap in and actually increase the temperature. However, in the case of radiant floors, the fluid leaves at 120° and comes back around 100°F–110°F. Solar works great with radiant floors.

If your existing heating system has been found to be incompatible with solar, that doesn't mean you can't use solar to heat the building. It just means you will have to use a non-integrated heat distribution method. The following sections describe both integrated and non-integrated heat delivery methods.

Integrating a Solar Heating System with a Forced-Air Heating System

It is possible to integrate a solar heating system with a forced-air heating system. This is commonly done in retrofit situations. If you are building a new home, however, we do not recommend a forced-air heating system, because the blowers use a lot of electricity to distribute heat throughout a building. People often neglect to include these electricity costs when they compare the efficiency of various heating systems, even though the electricity can add a significant amount to the total cost. For people who already have PV power, or are planning on having PV power, a forced-air heating system is not an alternative because of the large number of watt-hours it consumes. The only reason to ever have ductwork and blowers is for heat-recovery ventilation systems or fresh-air systems in commercial buildings. There are alternative heating systems with a much lower operating cost than forced-air systems. Although they might cost slightly more to install, their lower operating costs will quickly make up the difference. Ask your heating contractor about radiant floor heating, which is the best. There are other alternatives as well.

When integrating a solar heating system into a forced-air heating system, the objective is to get the heat from the storage tank, or directly from the solar collectors, into the house. The blower and ductwork associated with a forced-air heating system will do the job, and if they are already in use, they make the obvious choice. A liquid-to-air heat exchanger is placed in the ductwork near the furnace. It is always preferable to install the heating coil in the hot-air duct above the furnace. If an air conditioning coil is already installed in the ductwork, it will probably be in the hot-air duct just above the furnace. If there is room, try to place the heating coil below it. If there is no room below it, try to install it above the AC coil. We do not recommend installing the heating coil above the AC coil because the AC can freeze the heating coil during the air conditioning season by blowing freezing cold air through the heating coil that is filled with water. If there is no room in the hot-air plenum above the furnace, try to fit the heating coil in the cold-air plenum right next to the furnace. To do this you have to move the furnace air filter so it filters the air before it gets to the heating coil. If there is no air conditioner coil in the hot-air duct, and if central air conditioning will not be installed in this system, then install the heat exchanger in the hot-air duct just above the furnace.

Hot water from the solar storage system is circulated through the liquid-to-air heat exchanger (heating coil) and the furnace fan is engaged. As air is moved through the heat exchanger, it is warmed and then delivered throughout the building by the ductwork. In a system with no storage, the heat directly from the solar collectors is delivered to the heating coil; in a system with storage, the heat stored in the tank is delivered to the heating coil.

This heating circuit has several control functions that need to be addressed. For systems with no storage, the heat is delivered whenever it is available and needed by circulating the hot solar fluid to the heating coil and the engaging furnace blower. A summer heat diversion may be required and a summer/winter switch will be needed for the furnace–fan interface. In systems with storage, a two-stage thermostat or a low-voltage thermostat (exactly the same component that is used to turn the forced-air furnace on and off) will turn on the solar heating circuit when there is a call for heat and turn it off when the desired indoor temperature is reached. This thermostat should be set at least several degrees above the setting of the thermostat serving the conventional heating system. This way the solar system will do the maximum percentage of the heating load that it can.

This circuit should come on only if there is sufficient solar heat in the storage tank. Either a setpoint thermostat or an aquastat can be used to control this function. These devices measure the temperature of the storage tank and are preset to recognize the lowest temperature you want the heating circuit to operate at. For instance, we usually set the low temperature shutoff point at 85°F for a house where we like a temperature above 70°F. That is a difference of about 15 degrees. With this setting, when the tank temperature falls to below 85°F, the device will turn off the circuit and not allow it to come on.

A circulating pump must come on to circulate the hot water from the storage tank heat exchanger to the heat exchanger in the ductwork and back to the storage tank. This circulating pump is usually high voltage (120-volt AC), whereas the thermostat circuit is low voltage. A relay, which is wired into the thermostat circuit, will do this job.

The furnace fan must also be turned on when the solar heating system is engaged. The furnace has a terminal strip just inside the cover. There will be a row of small screws with letters by each one. These letters will include R, C and G. The R terminal provides 24 volts AC for the thermostat circuit. A thermostat wire is attached to the R terminal. This wire first goes to the thermostat. Then it goes to the setpoint thermostat (or aquastat). From there a branch goes to the pump relay. The other branch goes back to the furnace, where it is attached to the G terminal. The G terminal controls the blower. Here is how it all works: the temperature in the building falls to below the desired temperature. The thermostat calls for heat by closing a switch. Low-voltage electricity flows from the furnace (R terminal) through the thermostat to the setpoint thermostat (or aquastat). If the temperature in the storage tank is above our minimum setting, a switch will close in the setpoint thermostat and the electricity coming from the thermostat will flow through this switch. Some of this low-voltage power turns on the pump relay, which turns on the pump. The electrical energy then flows back to the furnace, where it goes to the G terminal, which engages the blower.

If there is a central air-conditioning system installed on this furnace, a single-pole double-throw switch must also be installed as a summer/winter mode isolation switch. This is because a central air-conditioning system also uses the G terminal to turn on the blower when it is engaged. When either the solar energy system or the air conditioner is turned on, the other system may turn on as well. When installing the switch, connect the solar lead to one side of the switch, and connect the air conditioning lead to the other side of the switch. A single wire is attached to the "out" pole of the switch and runs to the G terminal. The switch is flipped one way during the winter to allow the solar heating system to operate, and the other way during the summer to allow the air conditioning system to operate. This switch must be manually flipped at the beginning and end of the heating season.

Integrating a Solar Heating System into a Hydronic System

Retrofitting a solar heating system into a hydronic heating system is similar to integrating into a forced-air heating system. The basic concepts are the same except that instead of adding a heating coil to ductwork we are adding an interface to the heat delivery piping. As established earlier, integrating solar heat into low-temperature heating systems is practical and efficient, but integrating into high-temperature heating systems is not practical or efficient. Therefore, the following descriptions will revolve around low-temperature heat delivery systems.

Hydronic heating system integration is possible and sometimes fairly easy, but at other times it can be difficult and hard to figure out. Because there are so many variations in plumbing styles and methods, it is impossible for us to tell you exactly how to do your installation. If possible, seek the advice of the heating contractor who installed your existing system. If that contractor is no longer around, contact one who is willing to work with you on this project. If you want to do the whole installation yourself, you can learn a lot from studying the information at heatinghelp.com or other similar websites.

If you are installing this system in a new construction, we suggest you use a primary loop–secondary loop plumbing configuration. This configuration allows for multiple heat sources and multiple heat loads in a simply designed system. There are several good books available on this subject. Our favorite is *Pumping Away*, by Dan Holohan.

A main consideration when integrating a solar heating system with a conventional hydronic heating system is that you do not want to mingle the fluid in your solar storage tank or solar loop with the fluid in your hydronic heating system. The best way to integrate a solar system into a hydronic system is to use a large storage tank with a heat exchanger mounted inside the tank to extract the heat. The fluid from the hydronic heating system passes through the coiled heat exchanger so the fluids never mingle. If you are using multiple tanks, you must use an external heat exchanger or use tanks with dual heat exchangers inside the tanks to transfer the heat from the stored water to the heating fluid. If you are integrating a dump system, you must also use a heat exchanger.

To get heat from the solar storage tank, you will need an electrical and hydronic control system to divert the fluid when needed. There are several control options to choose from. In some cases you will be able to use the existing thermostats and controls in the home and simply divert the flow to the solar storage tank to pick up heat when it is hot enough. The main considerations here are the type of boiler that is installed and the operating temperature of the boiler loop. Some boilers are able to modulate their fire rate based on the demand. In other words, they are able to sense the temperature of the entering fluid and turn on high or low, based on demand and the set-point temperature. These are typically higher-efficiency units. If you are integrating into a system with a modulating boiler, you will be able to pre-heat the fluid before it gets to the boiler. Simply install a differential temperature controller to measure the temperature of the fluid returning from the heat delivery equipment and the temperature of the solar storage tank. When the tank is warmer than the fluid, the controller will engage a zone valve or motorized three-way valve to divert all flow through the solar storage tank heat exchanger. If the fluid is warmer than the boiler set point, it shouldn't have to turn on, but if the fluid is not quite warm enough, the boiler will finish the job.

Many low-temperature heat delivery methods use a high-temperature boiler (180°F) and temper the fluid down before it is delivered with the cold fluid returning from the heat delivery device. This is common in many manifolded or injection-loop systems or in systems that have a combination of both high- and low-temperature heat delivery methods. In these cases, you will need to bypass the boiler entirely by installing relays and switches.

You can also install a new two-stage thermostat or two separate thermostats in the building, one for the solar heating system and one for the backup. Set one of the solar thermostats or stage one of the two-stage thermostat slightly higher than the one for the backup system. If you have multiple zones, you will need a two-stage thermostat or two thermostats for each zone. As the temperature falls in the building, the first setting or thermostat to call for heat will be the solar thermostat, wanting to extract heat from the solar storage system and deliver that heat to the house.

The control circuit has to check the solar storage tank to see if there is any heat to deliver. A sensing device called an aquastat, or setpoint thermostat, does this job. A sensor is immersed in the storage tank and attached to the sensing device. The device can be adjusted to sense the minimum tank temperature that will provide hot enough fluid to work with the heat delivery system. If there is sufficient heat in the storage tank, a pump relay will turn the circulating pump on and heat will be delivered. Pump relays and aquastats are available from any hydronic heating supplier.

For systems without storage, it is most common to deliver the hot solar fluid directly to the radiant panel (floor, wall panel or ceiling panel). These systems typically have a collector array that is mounted to maximize the winter sun angle and minimize the summer sun angle. A diversion loop is also common. These systems heat only when the sun is out.

Non-Integrated Heat Delivery Methods

When retrofitting a solar heating system into an existing building, it is sometimes impossible to integrate it into the existing heating system. In this instance, the solar heating system will stand alone and work independently of a conventional heating system.

Options for delivering the solar heat into the building include radiant panels, a separate radiant floor system, one or more fan convectors, or baseboard units. With all these options, hot fluid is taken from the solar storage tank and circulated through the heat distribution device, where the heat is delivered into the building. The control circuit for any of these options is very simple, and the circuit is exactly the same for all of them. This simple control system consists of three basic components: a thermostat, a circulating pump and a setpoint thermostat (or aquastat).

This control circuit will operate just like the circuits described above for the integrated systems. The difference here is that you have the option of the control circuit operating at household voltage. The high-voltage method uses a line-voltage thermostat instead of a low-voltage thermostat. With this option, there is no low-voltage power supply, and all the wires must be sized for the high voltage. The advantage of this high-voltage system it that it eliminates the need for the low-voltage power supply and the pump relay. The disadvantage is that you have to run bigger wire to the thermostat.

Baseboard units are what you see in homes or buildings with traditional hydronic heating systems. They are composed of a copper pipe with radiating fins located along the length of the pipe. This finned pipe is covered by an attractive metal cover and is installed on the wall just above the floor. As warm fluid is circulated through the pipe, the fins get hot and heat the surrounding air. This warm air rises along the wall and, as it exits the heat exchanger, cool air is pulled in from the floor and is heated. This natural convection results in a very comfortable circulation pattern. Compared to a forced-air system, much less energy is required to distribute the heat throughout the building.

Although they are becoming rare, old-fashioned cast-iron radiators can work very well for solar dump-type heating systems. These large units can deliver a large amount of heat into a room efficiently. The downside is that they are big and take up a lot of room.

Fan convectors consist of a liquid-to-air heat exchanger with a fan to help circulate the heat. These units are often 18 to 24 inches tall, up to 10 inches deep, and can be 36 inches wide. You might be familiar with these as they are often used in motel rooms, usually located under the window. Fan convectors are especially useful when installing a solar heating system in an existing building because they are easy to install and do not take up very much room. Plus they can deliver a lot of heat very quickly. Fan convectors can be usefully combined with solar radiant floor heating systems for areas that might be hard to heat, such as bathrooms, which we like to keep warmer than other areas. Fan convectors use more energy to distribute heat into the dwelling than baseboards because they include a blower, which uses electricity.

High-Mass Systems

It is interesting to note that the first known use of radiant floor heating, a high-mass system, was by the Romans more than 2,000 years ago. Their famous bathhouses had a central heating system that heated the water used for bathing as well as for heating the building. A firebox was located below floor level of the bathhouse. The smoke and heat from the fire circulated

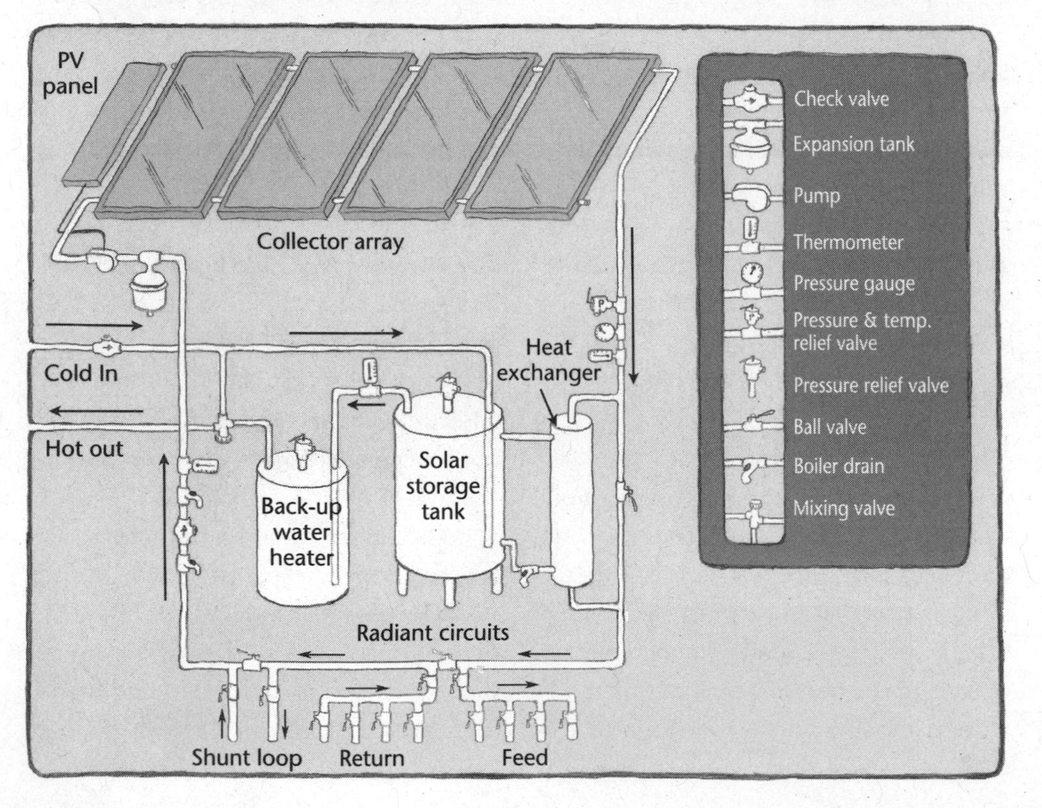

Figure 6.5: *Pressurized high-mass heating system*

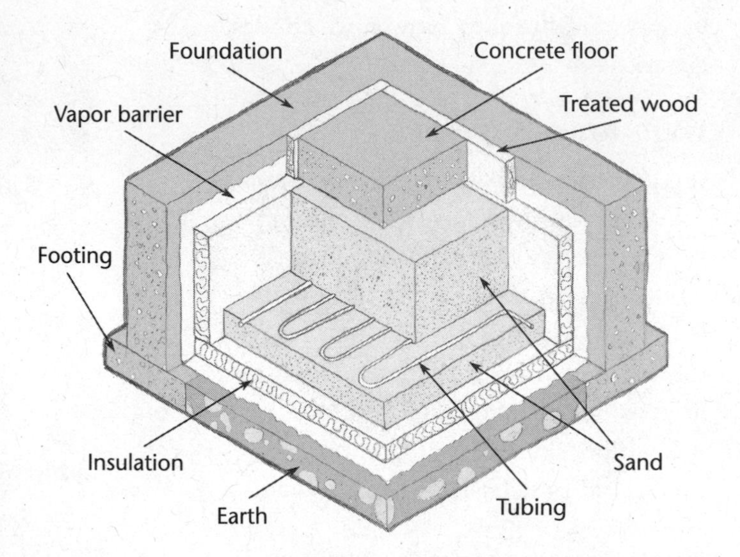

Figure 6.6: *High-mass system detail*

below the floor, heating it before exiting the building through numerous chimneys. These Roman bathhouses are known for their popularity and luxury. Way back then, people knew that radiant floors provided the most comfortable heat. Today we know that radiant floor heating is not only the most comfortable, it is also the most economical to operate.

General Description

A high-mass solar heating system uses a closed-loop antifreeze solar energy system to collect solar energy. This is exactly the same system that is used in a closed-loop solar heating system with storage. Instead of storing the heat in a tank of water, it is stored in sand under the floor of the building. The heat then slowly rises into the building, warming the floors and the whole building. If there is a domestic hot-water load in the building, a combination system can be configured so the system will heat the sand and the hot water.

This type of solar heating system uses a large amount of sand placed inside an insulated box located beneath the lowest floor of the building. The insulated box is made of two-inch extruded foam insulation.

We get the heat into the sand box using a gridwork of Pex tubing, similar to a traditional radiant floor heating system. The tubing is placed just above the bottom of the box, and hot solar fluid is circulated throughout the tubing, warming the sand.

Sand as a Heat-Storage Medium

The comment we hear most often about high-mass heat storage is that sand is a poor heat-storage medium and a poor conductor of heat. It is true that sand stores and conducts heat differently than water, but we can take those differences and use them to our advantage.

When a large amount of thermal mass is used, it can take a long time to heat up. Likewise, it takes a long time for this large thermal mass to cool down. We can use these traits to our advantage. This type of high-mass system can be turned on in mid to late summer. Because it takes a long time to heat the mass, we can begin to harvest solar heat well before the beginning of winter without overheating the building. We continue to charge the system with heat throughout the late summer and fall. In some climates, the system can run all year. Bob's system has never been turned off since it was commissioned in 2007. It is important to consider that the heat storage — the sand — never gets very hot. On the other hand, the sand bed is so large that it holds a large amount of heat. Because the sand-bed temperature is not much higher than the desired room temperature, the heat does not come out very quickly. There are times during the heating season that are very cold and the solar heating system cannot keep up with the heat loss of the building; at such times, backup heat is required to maintain the desired indoor temperature. Most buildings with high-mass systems do not have full-sized backup heating systems because they are rarely needed. The backup heating system has to add only a little heat during the coldest weather to finish what the solar can't. The backup heating system for a high-mass system cannot be a radiant floor system incorporated into the slab that is directly over the sand bed as this will interfere with the operation of the solar part of the sand bed. If radiant is to be used, a radiant wall or ceiling panel should be used. Another thing to consider when planning for a low-temperature heating system is the type of floor covering. Any insulation on the floor, such as carpeting, will significantly affect the performance of the system. There can be no carpet except for occasional throw rugs. Hardwood flooring has been used, but remember that anything on top of a radiant slab, except for tile, will affect its ability to deliver heat.

This type of system is classified as an unregulated system because there is no thermostat to control the indoor temperature. The fact is that high-mass solar heating systems rarely, if ever, overheat a building. One reason is the collector tilt angle. Collectors are tilted to maximize the low winter sun, so they are mounted at a steep angle. During the warm months of the year, the sun is high in the sky and therefore sheds little energy onto the collectors. In other words, the collectors self-regulate to a great extent by being very efficient at collecting heat during the heating season and very inefficient during the non-heating season.

For most locations, we recommend a sand bed that extends two feet deep under the whole building. Compacted sand weighs about 105 pounds per cubic foot. With a two-foot bed, our mass weighs 210 pounds per square foot of interior floor space. That equals 10.5 tons per 100 square feet of the building. For a modest

1,200-square-foot house, for example, the thermal mass would weigh 126 tons. This is a lot of mass. By comparison, the weight of a 350-gallon tank full of water is only about 3,000 pounds (1.5 tons).

Water is the medium to which all other materials are compared in terms of their heat-storage capacity. One cubic foot of sand can store the same amount of heat as 3.2 gallons of water. Water is about twice as dense as sand and transfers heat better because the air spaces surrounding each grain of sand inhibit the flow of heat. But even though sand may not be the perfect medium, it is cheap and easy, so using a lot is not a big deal.

By having a really big thermal mass, we can store a lot of heat. Depending on the overall efficiency of the building housing this solar energy system, we can usually store several months' worth of heat in our mass. In addition, whenever the sun does shine during the heating season, additional solar heat will flow into the mass, replenishing it.

People sometimes ask if a sand bed can be constructed outside near a building, with a layer of tubing placed at the bottom of the sand bed and another layer near the top of the sand bed. The bottom layer is used to deliver solar heat to the sand bed and the top layer is used to extract heat from the sand bed to be delivered to another location. We have not had any luck with this concept because the sand beds do not get hot enough to extract heat in this way, and sand transfers heat much differently than water. We have heard unsubstantiated reports of success in this area, but have no personal experience with it. We remain skeptical.

One of the reasons high-mass systems work so well is that they are located within the building envelope. This means that as long as the sand bed is warmer than the indoor air temperature, it will be transferring heat into the space. When the sand bed is moved outside, it requires a much higher temperature in order to get the heat into the building because the heat will have to be transferred to another fluid and another heat delivery system.

Living with a High-Mass Solar Heating System

Because every solar energy installation is different, a high-mass solar heating system will react slightly differently in every situation. The first thing to realize is that it may take an entire year to get the sand bed dry and hot. The time of year the system is installed also makes a big difference the first year. If the solar energy system comes online early in the summer and is allowed to operate most or all of the first summer, that system will perform relatively well the following winter. If the system comes online any other time of the year, the performance of the system will be diminished that first year.

If you look at the section in Chapter 7 about sizing solar heating systems, you find certain general sizing parameters. You will see that for a dump-type system, a certain amount of collector area is needed for a particular building. If we were to add storage to that same building, we would approximately double the amount of collector area. This additional collector area and storage allows us to heat the building for a longer period of time, usually for a whole day instead of a half day. You will also see that the sizing formulas developed for closed-loop solar heating systems with storage are basically the same as those for high-mass solar heating systems. Lastly, you will see that a high-mass solar heating system might provide 25 percent more heating than a closed-loop system with storage that has the same size of collector array. You are probably wondering how this can happen. The reason is that with a closed-loop solar heating system with storage, we turn the system on when the heating season begins in late September and turn it off when the heating season ends in May. In the case of high-mass solar heating systems, we turn the system on in early to mid-August and operate it through May. This earlier start date allows this high-mass system to collect nearly two additional months' worth of solar energy using the same collector area, thus increasing the overall efficiency of the system. Note that these additional days of operation come during a traditionally sunny time of year, so we harvest even more solar energy. The large thermal mass allows us to store that energy for later use.

Once your high-mass solar heating system is installed, you will have to develop your own particular operating schedule. The earlier in the summer that you can turn on your system, the better. You will have to experiment. You might find that during the first year or two of operation, you have to turn on the system early in August. After that you might delay the turn-on date till later in August. But if it is cool at night, most people just leave the windows open and cool the building with fresh air. Even with the windows open and the building at a comfortable temperature, the mass is still heating up. The warmer we can get that mass during the sunny months preceding winter, the better. Most people with high-mass solar heating systems are able to keep their windows open much later in the early winter than those using conventional heating. This fresh air is a real bonus for those with high-mass systems, especially in locations such as Wisconsin, where most buildings are closed up tight for more than half the year.

In the late spring, you will want to turn off the system once the building is consistently too warm. The timing of this will depend on your location and how much sunny weather you are experiencing that particular spring. You might find that during the spring you have the system off for a week and then back on for a week till the temperatures outside become uniformly warm. Of course, the solar water heating portion of the system will continue to operate all year long.

Another advantage of high-mass systems is that dehumidification requirements are eliminated. If the system is in a basement, the basement will be warm enough to eliminate condensation and the resulting mold, etc., that goes along with sweating basement walls. Where the sand bed is on grade, the floor should be kept a little warm all summer to eliminate condensation. This will not overheat a house or interfere with air conditioning.

Air-Type Solar Heating Systems

One of the most cost-effective solar heating systems is an air-based system. This is because the equipment is simple and inexpensive. Air is the heat transfer medium instead of solar fluid. Ductwork is used to transfer the heat from the collector to the building. These systems are most popular in climates that have large heating loads. On cold, sunny winter days, air systems can produce impressive amounts of heat, especially if there is snow on the ground. Air systems are strictly space heating systems and are not used for water heating. We will get into that a little later in this section. Air collectors are usually vertically mounted on a wall and can have the same Btu output as a similarly sized liquid collector.

Air-type solar heating systems suffer from the same flaw as conventional forced-air systems. A substantial blower is required to distribute the heat collected in the collector through the building. This electrical load found in all older models can account for up to half of the total electrical demand of an electrically efficient home during the heating season. Air-type collectors are less expensive per square foot to purchase than liquid-type collectors.

Although these systems are cost effective, they do have their limitations. The biggest is that they can be set up only as systems without storage. Back in the 1980s, these systems were set up with heat-storage components of many kinds: rock bins, closets full of bricks, special tubes filled with eutectic salts, and other apparatus. Hot air was circulated through these kinds of thermal mass, and the mass soaked up the heat and stored it for use later. To get the heat out, cool air was circulated through the mass, where it picked up any available heat that was stored there. Most of these systems worked for a while, but problems soon arose. The most significant problem involved air quality. As dust and dampness accumulated on the mass material, a medium was created for the growth of mold and other nasty things. These pollutants then were circulated into the building, creating potential health problems. Rock bins also rotted, leaving a huge mess. Eutectic salts, which are very corrosive, leaked out of their containers. Do not attempt to have a dedicated heat-storage setup for an air system.

When set up without storage, air systems can be a good choice for solar heating systems, especially in commercial applications. Two versions of air-type heating systems are used today. The most common type takes cool air from inside the building and circulates it through the collector, where the air is heated. The warmed air is then delivered back into the building. Back in the early and mid-1980s, I manufactured and installed many dump-type air systems. Almost all those systems are still working fine today and saving homeowners money. These systems consisted of a wall-mounted collector and minimum ductwork with a blower and controller. They were simple and inexpensive and were designed to provide about 25 percent of the annual space heating needs.

Today, some of the most popular applications of air collectors are in warehouse facilities, shops and workplaces. These facilities usually can tolerate some modest temperature swings on a daily basis and also usually are filled with mass that can soak up solar heat during the daytime, when it is plentiful, and slowly give off that heat during the night.

Many people also enjoy warmer temperatures in their homes during sunny days. Bob often recalls one family in particular. They were a family of five; both adults were schoolteachers. Bob installed an air system on their home during the week and was done on Friday. That Saturday about noon he got a call from them. It was a perfect sunny winter day with outside temperatures hovering around 10°F. It seems that these folks were quite conservative with their thermostat settings, so their house had not been very warm during the winter months. The call was to invite him to a spontaneous Bermuda shorts party because the solar heating system was making the house warmer than it had ever been during the winter. Needless to say, they were happy.

The less-common type of air system uses unglazed transpired collectors that draw fresh air through the collector, where the air is heated and then delivered into the building. These systems are designed for buildings that require large amounts of makeup air. Because these systems are for space heating only, they are often wall-mounted. Wall-mounted systems work well during the winter, when the sun is low in the sky; they don't work during the summer, when the sun is high in the sky. They also work best in buildings that contain a large amount of thermal mass, such as warehouses and shops. Because the solar fluid in these systems is air, and ductwork is required to move that solar fluid within the system, most air systems are building-mounted. Running ductwork outside is not very efficient.

Air System Controls and Additional Components

It is possible to design PV-direct systems that use a PV collector wired directly to a DC blower. This design is most appropriate for small systems that are wall-mounted. As this setup has no controller or thermostat, there is no overheating control. Also, DC blowers are hard to come by.

We like to use differential controllers with air systems. Put one sensor on the absorber plate in the collector and the other in a central location within the heated building. This configuration allows the system to turn on as soon as there is available heat within the collector, and it shuts off the system when the building starts to get overheated. This is especially efficient when heating a building that is kept cool during the evening. In this case, the collector can come on earlier in the day. In addition, the temperature differential between the building and the collector is higher, thus increasing collector efficiency. A high-limit thermostat and relay can also be added for more precise overheating protection. With this option, when the thermostat reaches high limit, the power is cut off to the blower. Note that it does not hurt an air-type collector to turn it off and allow it to stagnate.

Another way of regulating the fan in air systems involves turning the collector on at a preset temperature, using some type of setpoint thermostat. This type of control is the least efficient of any discussed here.

Another very important component in an air-type heating system is a set of dampers. We learned earlier that water thermosiphons within a tank. Well, air can thermosiphon within an air system and even within an air duct. This phenomenon is most apparent with roof-mounted collectors, but it also happens with wall-mounted collectors. At night, when the temperature drops, the air within the collector also gets cold. Because cold air is heavier than warm air, it can drop down the ductwork and flow into the building, and at the same time warm air from the building can rise to the collectors and lose its heat. To prevent this, some type of damper must be installed on each primary duct going to and from the collector. Motorized dampers are used and are set on the same electrical circuit as the blower. These dampers must have a positive seal or heat loss will occur. Spring-loaded dampers, where the blower creates enough pressure to open the damper, have also been used. The only exception to the above damper rule is for wall-mounted PV-powered systems. In this case, it may be possible to use a flapper damper that is placed on the lower cold-air duct. These dampers can be made using hardware cloth and ripstop nylon. This design also requires that the cold-air inlet be rectangular and horizontal, and that the hot-air duct be located at the top of the collector.

Heating Water with Air Collectors

Many homeowners and installers have attempted to heat water with air systems, especially during the 1980s. This system incorporated an air-to-liquid heat exchanger in the hot-air duct, a water-storage tank and a circulating pump. The circulating pump came on whenever the blower came on. As hot air passed through the heat exchanger, the water theoretically would be heated and the pump circulated the water from the storage tank through the exchanger. The overall efficiency of this system is very low. The air temperature is never very high, so the temperature differential between the air and water is small. This does not lead to very good heat transfer efficiency. Also, if you add up the electrical energy used to run the blower and the pump and the dampers, you don't come out very much ahead. But the biggest issue has been freezing heat exchangers. The heat exchanger must be located on the warm side of the damper. If the damper ever fails during freezing conditions, the heat exchanger will freeze and burst. Every water heating system we have seen that used air collectors has frozen at least once. In the worst-case scenario, this is an especially ugly failure because when the exchanger thaws out, water shoots out of the rupture; and this water is hooked up to your main water supply. Say the exchanger freezes overnight and you don't know it. Everyone leaves the house for the day. The exchanger thaws out and water squirts out of the exchanger all day long. You come home to a flood and a big water bill. Heating water with air collectors is a bad idea. If you insist on trying to heat water with air collectors, we suggest you add a heat exchanger to the storage tank and fill the heating loop with glycol. This may require two pumps if an external heat exchanger is used or just one pump if an in-tank heat exchanger is used.

Additional Comments on Air Systems

A few specialty applications of air systems worth mentioning involve solar chimneys.

A solar chimney is a vertical duct that is glazed on the south side and black on the inside. The duct is open on the bottom and the top. This duct is actually a solar air collector. When the sun shines on the collector/duct, the air inside is heated and rises. As the air rises and leaves through the open top, it pulls fresh air into the bottom. Air circulation is created. When these collectors have been used for cooling, especially in warm climates, they are incorporated into the design of the building. The bottom of the duct is open to the building. As air rises in the duct on a sunny day, air is pulled out of the building. The air can be taken off the ceiling, where the building is hottest. Cooler air can be drawn into the building by the negative pressure caused by the solar chimney.

A promising technology that we have hopes for is the use of air rising in a solar chimney to spin an electric generator. Massive solar chimneys in desert areas could potentially generate significant amounts of power. Some of the proposed designs have the solar chimney going up the side of a large hill or mountain. For current developments in this area, just enter "solar chimney" in your Internet search engine.

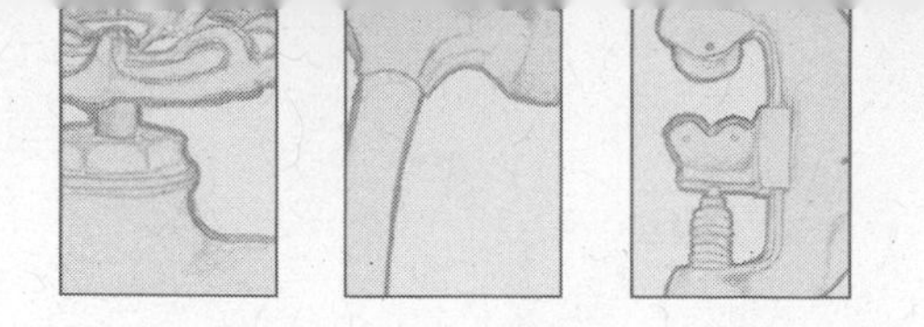

7

CHOOSING THE RIGHT SYSTEM FOR YOUR NEEDS

AFTER YOU HAVE PUT IN PLACE your energy conservation measures, you are ready to decide what type of solar water heater will be best for you. There are several types of solar water heaters to choose from. Each type is slightly different and each is appropriate for different situations. However, as stated earlier, we recommend only the ICS, drainback and pressurized systems. These three systems should cover all possible climates.

When selecting your system, we caution you to always plan for the worst-case scenario. Situations have a way of constantly changing, and what might be okay now may not be in a couple of years. Remember that a properly installed solar water heater usually lasts decades. Think ahead. While we are talking about worst-case scenarios, let's learn something from the past. You might recall from our review of the history of solar energy that, in California's first solar boom, some people had bad experiences with their solar water heaters because they purchased inferior and untested products. In most cases, they were sold on low system cost. Also recall that during the Florida solar boom the same thing happened for the same reasons. We experienced a similar case during the tax credit days of the 1980s. Most systems were good, but there were, and always will be, people out there trying to make a quick buck selling inferior products. This always seems to be the case when something becomes fashionable.

What can you do to avoid this experience? First, use your common sense. If a deal is just too good to be true, then it probably isn't good or true. (There are no free lunches out there.) Study this book to learn what it takes to make a quality system. Talk to people with experience. Be careful. If someone is trying to sell you something and they say that you must make the decision today (usually right after a sales pitch), don't do it.

We have found that 90 percent of the problems encountered with solar water heaters result from wrong (usually inferior) components, poor workmanship or an inappropriate system for your climate. Of all the solar water and space heating systems we have installed over the past 30 years, only about 10 percent of them have ever required a service call.

Before selecting a system, you may want to find out if the rebate programs in your area require specific types of systems or components. Most utility rebates here in Wisconsin, for example, require that you install either a pressurized or drainback system because of the climate. Many rebate programs also require that you use certified and rated collectors to qualify. If you are dealing with a company that is new to the business, it would be wise to specify a certified system. Experienced installers may offer a high-quality system that is not certified; but certification is a wise investment anyway. Be sure to carefully check the credentials of the people you are dealing with.

Finally, the system you select may be determined by the installation limitations of your site. For instance, a drainback system requires that the collectors are mounted above the drainback tank. If your only option for a collector array is on the ground, then a drainback system will not work for you. The following sections will help you choose an appropriate location for your collectors and help you size the system according to your needs.

Siting a Solar Energy System

Sometimes it seems silly to state the obvious, but here goes: solar collectors must be in the full, direct sun if they are going to work properly. This may seem obvious, but you wouldn't believe how many times we've had prospective customers tell us there is lots of sun where they are thinking of placing the collectors, only to find the spot in considerable shade. So, what is the bottom line for choosing a location for solar collectors? Read on.

First, let's review how the sun shines on the Earth. The relative movement of the sun across the sky determines our days as well as our seasons. The path of the sun changes every day. Here in Wisconsin, on the first day of spring (March 21) the sun rises directly in the east; at noon it is directly to the south and is above the horizon at about a 45 degree angle, and it sets almost directly west. On that day, the sun is up for 12 hours and down for 12 hours. On the longest day of the year, the first day of summer (June 21), the sun rises in the northeastern sky, is directly overhead at noon (68 degrees from horizontal) and sets in the northwest. The first day of fall (September 21) is exactly like the first day of spring. The shortest day of the year is the first day of winter (December 21), and the sun rises in the southeastern sky, is directly to the south at noon and is about 23 degrees above horizontal (low in the sky), and then sets in the southwest. The height of the sun, in degrees, will vary depending on your geographical location. The further north you go, the lower the sun will be during the winter months.

Generally, solar collectors should face within 30 degrees of south, be mounted at an angle to the sun that will maximize their performance, and be in the direct sun (no shading at all) from 9 a.m. until 3 p.m. It is between these hours that a fixed point will receive 80–90 percent of all the solar radiation it receives over the whole day. Some solar installers advocate for full sun only from 10 a.m. until 2 p.m. In some cases this will be suitable, but for optimal collection, you should try to have full sun between 9 a.m. and 3 p.m.

To get the maximum performance from your solar energy system, the collectors must face the sun at noon. Both early and later in the day, the energy coming from the sun must pass through more of the atmosphere than during the middle of the day. This atmosphere is full of dust, water vapor and moving molecules. All this stuff in the air interrupts and weakens the energy flow from the sun to the Earth's surface. The less atmosphere the sun's energy has to pass through, the stronger it will be. It may seem elementary, but solar collectors need to be in the full sun for most of the day. They will not work if the direct sun must pass through trees or other vegetation, even if the leaves are gone. A tree without leaves will block up to 75 percent of the sun's energy.

Collectors should face as close to true south as possible, but a variation of up to 30 degrees is generally acceptable and will not significantly reduce panel performance. To find south, you first have to find north and then look the other way. However, compared to a compass reading of north, true north is not the same everywhere. To find true north at your particular location, drive a stake vertically into the ground and watch its shadow. When the shadow reaches its shortest length, it is pointing exactly north. Unless you have time to sit around staring at a stake in the ground, you probably want an easier way to find true south. You can also find the declination of the compass reading for your area from your local weather service, or by consulting

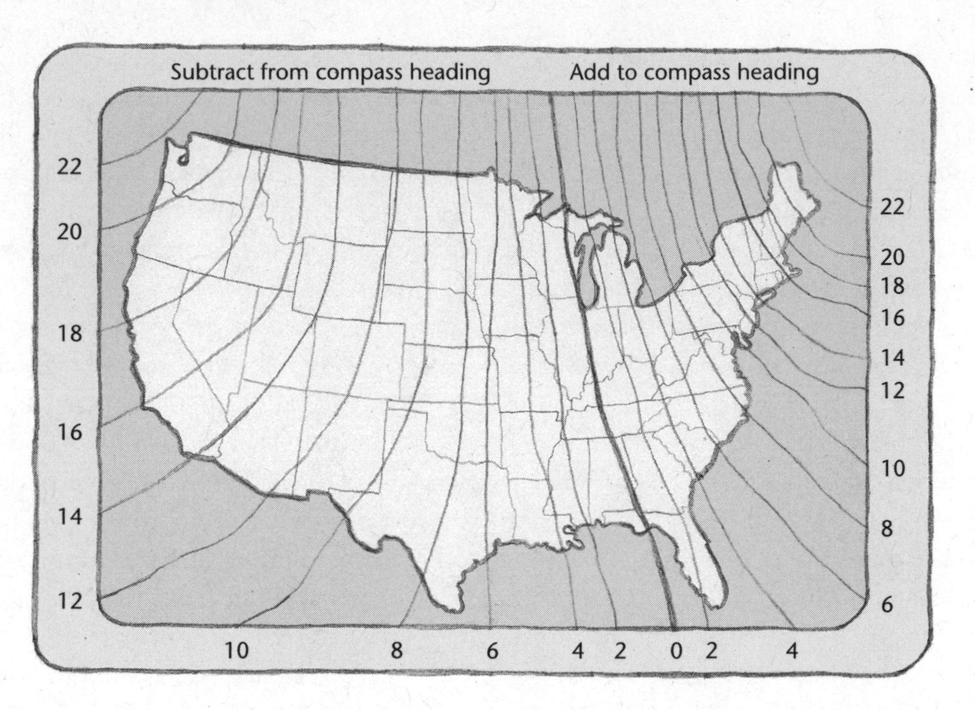

Figure 7.1: *Isogonic map of the United States*

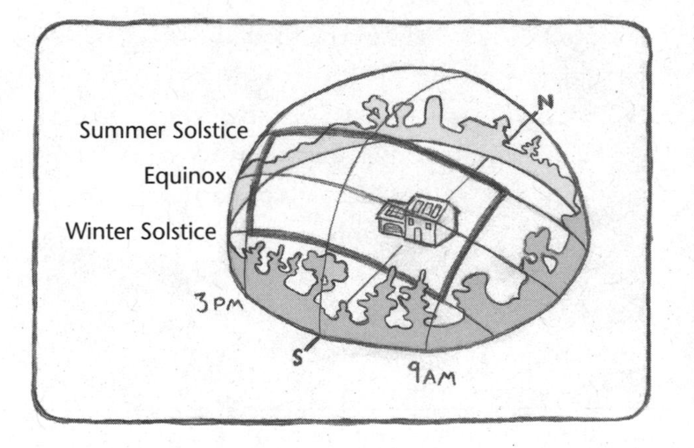

Figure 7.2: *The solar window from above*

an isogonic chart. An isogonic chart demonstrates how the compass reading of north will vary depending on global location. To use the isogonic map in Figure 7.1, you simply find your location on the map and note the nearest line. Those on the western half of the United States will have to subtract the given number of degrees from their compass reading and those in the eastern half will have to add them. For instance, if you live in southern California, you will need to subtract between 14–16 degrees from what your compass reads as south. Many compasses have a dial for you to make this adjustment.

One way to determine whether a proposed site is good is to imagine a big window in the sky directly to the south of your collector array. Imagine that all the sun's energy that will fall on your collectors must pass through this window. The

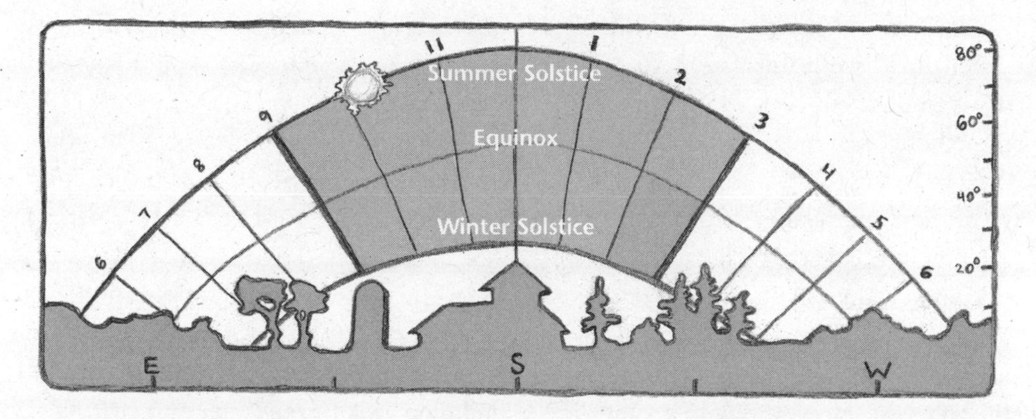

Figure 7.3: *The solar window from ground level*

solar window is defined by the path of the sun across the sky throughout the year. On June 21, the summer solstice, the sun takes its highest path of the year across the sky. The lowest path throughout the year is on the winter solstice, December 21. These paths of the sun define the top and the bottom of the window. The sides of the window are defined by where the sun is at 9 a.m. and 3 p.m. You want this solar window to be completely open and free of any obstructions.

Professionals may wish to purchase an instrument that can be set up at any location to show the solar window. There are several on the market today that work successfully.

The first, the Solar Pathfinder, uses a plastic dome to reflect the solar window onto a sun chart. The sun path chart is selected based on your latitude and will have the appropriate sun angle for the different months of the year. You trace the shadows reflected on the dome onto the sun chart, giving you a concrete analysis of any obstructions and where they are. The Solar Pathfinder will also allow you to rotate the surface to account for magnetic declination, which is a handy feature. This tool also has software in which you can upload a digital photograph of the reading in order to determine the percentage of total annual solar radiation that the proposed site will receive. There are two very important considerations when using site assessment software to analyze solar thermal performance: first, are you able to set the hours that will be analyzed and second, is all shading considered 100 percent or can it be adjusted? All site assessment tools currently on the market were originally designed to assess a site for PV. As such, they typically analyze the dawn-to-dusk solar window because PV is typically put on a tracker where the collectors move to face the sun directly at all times. Because solar thermal collectors never track, we do not need to analyze dawn-to-dusk but need to analyze only the three hours before and after solar noon. Analyzing anything more than that will give a false impression of the output of a solar thermal collector. The other big consideration is the way shading is considered in the analysis. As you may know, any shading on a PV panel can completely turn off the collector. However, with a solar thermal collector, a little shading will not turn off the whole collector; only the shaded part will lose productivity. Most software analyses consider any shading at 100 percent, but for solar thermal it probably is not 100 percent in most cases. Consider deciduous trees when they lose their leaves during the winter. On average, 50 percent of the sunlight is blocked and refracted, and the other 50 percent shines through, so shading behind a tree with no leaves is typically 50 percent shading, not 100 percent shading.

Figure 7.4: *Solar Pathfinder*

BENJAMIN NUSZ

The Solar Site Selector is another option. It is composed of a semicircular base in which you attach a transparent sun path chart. This sun chart will vary based on your geographical location. To use the Solar Site Selector, point the instrument straight south, keep it level and then look through an eyepiece located on the base. When you look through the eyepiece you will see the sun chart transposed over the actual solar window in the sky. The advantage of this instrument over the Solar Pathfinder is that you can actually see any potential obstruction instead of just its reflection. However, you do not end up with a concrete recording of the solar window.

If you are not willing to purchase one of these instruments, or you just want a quick way to determine obstructions, you just need to find out the angle of the sun during the winter solstice for your area. Figure 7.4 lists the approximate angles of the sun on December 21 at various latitudes at different times. Next, take a simple plastic protractor and a piece of straight wire. Bend the end of the wire and insert the bent end into the hole in the base of the protractor. Hold the protractor so that the flat edge is up, allowing the wire to point down.

Stand where you plan to locate your collectors and tilt the protractor so that the wire is pointing to the appropriate angle for your location for the appropriate time. Protractors typically have the 90° mark at the tip of the circle, so you will have to subtract your sun angle from the 90° to get the "protractor angle." Next, point the protractor to where the sun will be during that time of day. At noon, the sun will be straight south. At 9 a.m. the sun will be just under 45° east, and at 10 a.m. the sun will be just under 30° east. Look along the straight edge of the protractor. If you see anything other than sky, you will have obstruction at that location during the winter solstice. Determining 2 p.m. or 3 p.m. should be done using the same angles. Now, this is a rather crude method for determining obstructions, but it is simple enough that you can do it without specialized tools.

If using any of the solar window assessment tools above reveals that more than 10 percent of the solar window between 9 a.m. and 3 p.m. has obstructions of any kind, you might want to re-evaluate the location. If the obstructions are tree branches, you may be able

Figure 7.5: *Using a Solar Site Selector*

BOB RAMLOW

Figure 7.6: *Sun angle chart during winter solstice*

	Time of Day (During Winter Solstice) 9AM	10AM	12PM	2PM	3PM
Latitude North	**Approximate Sun Angle (degrees of solar altitude)**				
28 degrees	23	31	39	31	23
32	20	28	35	28	20
36	17	24	31	24	17
40	14	21	27	21	14
44	11	17	23	17	11
48	8	14	19	14	8

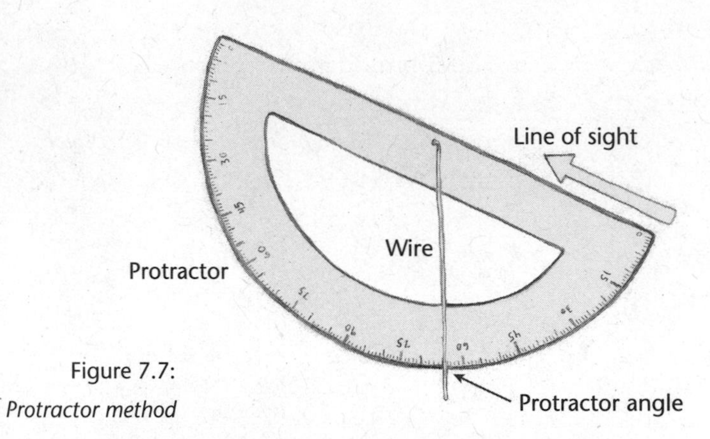

Figure 7.7: *Protractor method*

to remove them. However, remember that trees grow, collectors don't. If the obstructions are only at the very bottom of the solar window, then the shadows will be cast only when the sun is lowest and the days are the shortest. This is the time of year when the solar energy system will produce the fewest Btus per day, so you may be willing to sacrifice some performance. This is a judgment call you will have to make. We tend to go for no shading. If the location where you would like to have collectors has just a few obstructions, it doesn't mean that solar is not for you. Sometimes you may be able to increase the size of your collector array to accommodate for shading. We always want the most out of every collector we install, so we tend to be kind of sticklers.

Many locations can be suitable for mounting collectors. Collectors do not always need to be mounted on a south-facing roof or even on the roof at all. Figure 7.8 displays many of the collector mounting locations that have been used successfully. You also have the option of mounting the collectors on a rack on the ground. However, these are not the only options: any place will work as long as it has a clear solar window and you adhere to the appropriate mounting methods.

Roof-mounted collectors are more likely to be in a sunny location than ground-mounted collectors because they are higher in the air. As you evaluate your site for the best location for the collectors, you will see that the higher you go, the less shading you will encounter. Roof-mounted collectors also take up less room in your yard than ground-mounted collectors. Another advantage of roof mounting is that the length of piping required is often less than for ground mounts. Roof mounts are also less likely to be affected by vandalism than ground-mounted collectors. On the other hand, sometimes the dwelling is located in a shady spot and you would rather not cut trees around the house just for the collectors. In this case, you may be able to install a ground mount near the dwelling, in a better location to collect direct solar radiation.

This was the case at Bob's first home. It has several large maple trees in the yard directly south of the house. Those trees

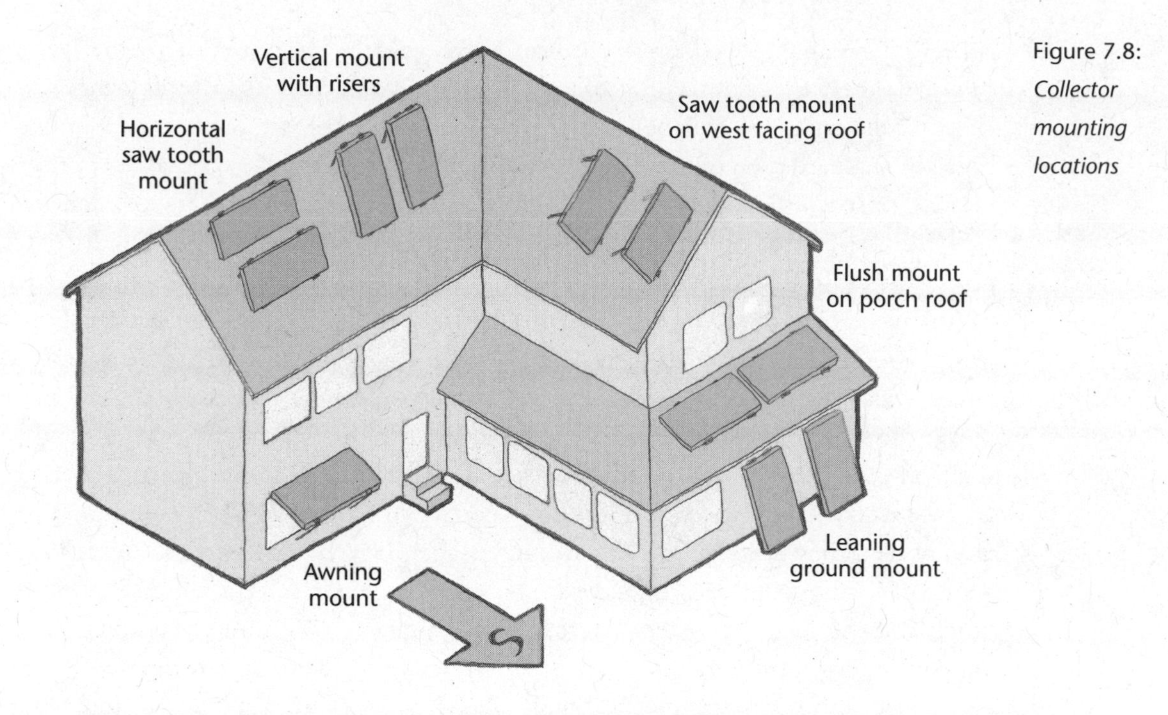

Figure 7.8: *Collector mounting locations*

keep the yard and home nice and cool all summer and are also a haven for many birds. Consequently, the collectors were ground-mounted about 75 feet from the house and away from the trees. A ground-mounted system often costs slightly more than a roof-mounted system because of the added cost of the rack that holds the collectors, as well as the added costs involved in trenching for the pipe runs. You will want to find the best location for your site. If there are multiple locations, all with clear solar windows, then you should consider other variables, such as aesthetics, cost, ease of installation, length of pipe runs and/or weight load.

Sizing a Solar Water Heating System

Solar water heaters have been around for a long time. Over all those years of experience, some general sizing rules have emerged. We have found that using a simple sizing method in designing solar domestic water heating systems has worked very well. However, you will want to adjust the rules to meet your specific needs. There will also be times when the simple rules do not apply and you need to go through a series of calculations. This is particularly true in commercial applications. Often, these calculations will also need to be done when you are using evacuated tube collectors (because collector performance can vary widely between manufacturers). In the following, we will demonstrate both the standard rules of thumb and the method of specific calculations.

It should first be noted that in most cases you can't size the system to heat 100 percent of the water you use. You probably don't live in paradise, so you are going to have cloudy days. You will need to size the system to cover 100 percent of the load on a clear sunny day. If the system were sized to heat everything on a cloudy day, then when you did have a sunny day you would be producing too much and overheat the system. Here in Wisconsin, a system sized to cover the entire load during the summer will typically cover 50 percent of the load in the winter and will typically heat about 75 percent of an annual load. Locations with milder winters can usually reach a higher annual percentage.

Calculating the Load

To begin sizing your system you will want to start with the load and then work backwards. To determine how much water you need to heat, you need to know how much you use. On average, each person in a modern household uses 20 gallons of hot water per day. If you simply multiply the number of members in a household by 20 gallons, you will have your household's average daily hot water load. However, a number of factors will alter this estimate. As the number of members in a household increases, the average usage per person will decrease. This is due to economies of scale. For instance, you are more likely to wash full loads when you have many to do and people tend to take shorter showers if there is a line. In other words, a family of five typically uses 10 to 20 percent less hot water per person than a family of two or three. Are you average or are you conservative in your hot water usage? If you have been successful in reducing your hot water consumption through conservation techniques, then you should reduce the estimate of your load. If you feel that you use more hot water than average, you will need to increase your estimate. For instance, if you have a spa tub that holds 40–50 gallons of water and you use the spa regularly, you have to take that into consideration. Temper the average with consideration for your water consumption habits.

Simply using the average is the quick and easy way of sizing. As we said before, this is usually good enough for most cases. However, if you think that your load is much different than most and you want to find out what it is specifically, or if you are designing a system for a commercial application, then you will have to put forth a little more effort. The most accurate method for determining how much hot water you use daily is to actually meter your current hot water consumption. For most residences, this would be an expensive test. To do this measurement, you would have to install a water meter on the hot water line of your home, run the test for a month, and then divide total consumption by the number of days in the test to get an accurate average daily hot water consumption figure. Most likely this is not an option. However, with a little clever investigation you can usually get a pretty close estimate of your average load without a water meter.

The best place to start is with your utility bills. If you use a natural gas water heater, look to see what your usage was for the month of July. Unless you have a pool heater or a gas range, you used gas only in the water heater, so you will know the number of therms you use on a monthly basis. A gas range will actually use less than a therm a month, so it usually isn't a big factor. In addition, we have a tendency to use less hot water in the summer than in the winter, so you will want to increase July's bill by five to ten percent to get your average monthly consumption. You can't determine the average just by looking a monthly bill if you use propane to heat your water. You don't get a monthly bill. This would work only if you had your tank filled once the heating season was over and before the next heating season began. Then you could take the average over that time period.

Dish Washing	**Compact**	**Heavy Duty/Large**	
Energy Saving Cycle	5 gal	10	
Heavy Wash Cycle	10	15	
Pots and Pans Cycle	15	20	
Manual Dish Washing	4-8		
Clothes Washing	**Front Loading**	**Top Loading**	**Heavy Duty/Large**
Warm Wash/Cold Rinse	5	10	12
Hot Wash/Cold Rinse	9	19	24
Hot Wash/Warm Rinse	14	28	36
Hot Wash/Hot Rinse	20	38	48
Personal Hygiene			
Wet Shaving	2-5		
Daily Washing	2-5		
Tub Bathing	15-35		
Showering	2-4 gallons/minute		
Food Preparation	2		

Figure 7.9: *Average consumption chart in gallons of hot water*

If you have an electric water heater, you will not be able to determine your average consumption by your utility bills. You use too many other appliances to be able to isolate just the water heater. However, you can use a kWh meter. Usually, you simply plug the kWh meter into your outlet and then plug the appliance into the meter. Most electric water heaters are directly wired, so they will not have the plug necessary for most meters. If this is the case, you will want to ask an electrician for assistance. You don't always need to purchase the kWh meter, although it is rather handy in reducing your electrical consumption. Many libraries have one that you can check out for a week or two and sometimes you can borrow one from your utility. We suggest using the meter for an entire month, if possible, to find out an accurate daily average. Using it for only one day usually doesn't work because you may or may not wash the dishes or a load of laundry that day. These can be tracked as weekly tasks and averaged in to your daily load.

If the above methods don't fit your situation, you can still accurately estimate your average daily hot water consumption. To do this, you need to analyze the hot water consumption of each person in the household. Most people tend to underestimate the amount of hot water they use, so be careful to examine each person's daily habits carefully. To help you determine hot water consumption, the information in Figure 7.9 may be helpful. Using this method is usually time consuming because you will need to take into account the flow rate of your faucets and the efficiency of all of your appliances. The numbers in the chart are just averages, so if you are looking for specifics, you will have to do all of the hard work. To save time, we suggest just using the average of 20 gallons per person described above.

Sizing the Solar Storage Tank

Once you have a general impression of your hot water load, you can start to size the system components. You will want to start by sizing the solar storage tank. As a general rule, for every gallon of hot water you consume daily, you will want to have one gallon of solar storage. This rule is straightforward because the sun comes up daily and will be able to replenish what you use. An average family of three will probably want to have a 60-gallon storage tank to accommodate their 60-gallon-per-day load. Tanks come in a limited range of sizes, so you will have to select the one nearest to your need. We recommend rounding up instead of down. It is okay to have a little bit too much storage, but if you have too little, you risk overheating the system.

Sizing the Collector Array

Knowing how many gallons you need to have in your storage tank will allow you to size a collector array that will heat that amount of water. All you need to determine is the ratio of the number of square feet of collector to the number of gallons of solar storage. The ratio of collector area to storage capacity will vary depending on your geographical location because of the variance in the amount of solar radiation. Simply put, the more sun a location receives, the more water you can heat with a square foot of collector. Figure 7.10 shows generalized solar zones of the United States. This will help you determine the amount of collector area you will need for each gallon of storage.

Gallons of storage per square foot of collector:

Zone 1: 1½–2 gallons
Zone 2: 1¼–1¾ gallons
Zone 3: 1–1¼ gallons
Zone 4: ¾–1 gallon

Now you can appropriately size your collector array based on these ratios. For instance, using the ratio of 1 square foot of collector for every 1½ to 2 gallons of solar storage, a family of four in Arizona would want about 80 gallons of storage and 40–50 square feet of collector. Simply divide the storage capacity by the ratio number for your appropriate zone. If the same family lived in Wisconsin, they would want 70–80 square feet of collector. The zones are rather wide, so there will still be a range within each one. Generally, the further north you go in a zone, the more collector area you will want per gallon. The chart can be used to summarize the simple sizing method. It assumes that the smallest storage tank you can find is 40 gallons, so the numbers for one person are the same as for two. Forty gallons is the smallest size you would want to install, anyway. You also need to remember that flat plate collectors come in only a few standard sizes. Most common are 4' × 8' and 4' × 10'. However, you can often find them in 4' × 6' or 3' × 8' or some other odd sizes.

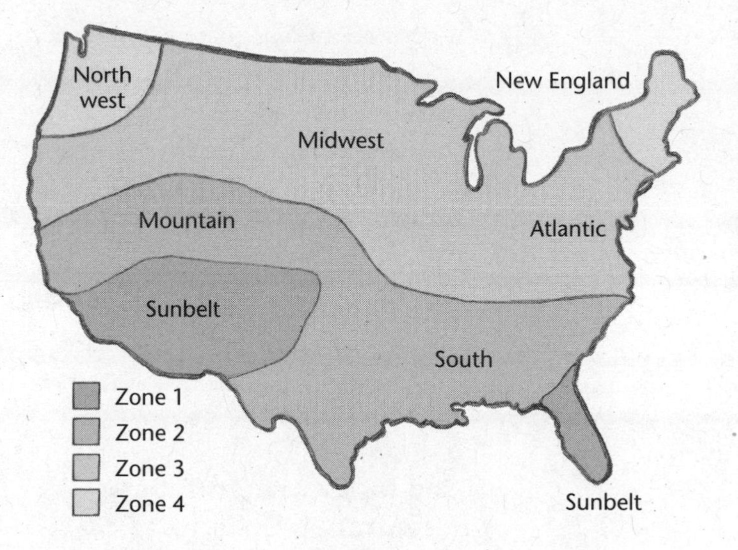

Figure 7.10: *Storage zone map*

You will need to design the system using these limited sizes. For instance, if you need about 51 square feet of collector, you should use two 4' × 8' collectors to come up with 48 square feet. As long as you don't go too far over the upper limit of your zone, you should be okay.

Please remember that the simple sizing method uses averages. Also remember that a little more or less collector area will barely be noticeable because solar radiation is variable. If you are in doubt, always upgrade to the next larger size.

Sizing for Evacuated Tube or Concentrating Collectors

Most evacuated tube and concentrating collector manufacturers will include sizing instructions in the manual. They will let you know the number of evacuated tubes or the size of the concentrating collector you will need based on your load and your geographical location. However, if this is not provided, you can determine the size of the array if you know the efficiency of the collector. You will want to know the number of Btu per square foot per day that the collector will produce for your location. This can usually be found through SRCC. The average for most evacuated tube collectors is about 600–700 Btu/ft^2/day. Concentrating collectors have much more variance.

Next you would want to convert your daily hot water load into daily Btu load. Instructions on how to make this conversion are covered more clearly in the section on calculating the space heating load. Take the total load and divide by Btu/ft^2/day of your collector, and you will get the number of square feet you will need. You can also use this method for flat plate collectors. Selecting the correct size of collector is much more vital for systems that use evacuated tube or concentrating collectors. They should be matched carefully to both the size of the storage tank and the load. With either type, too much collector or not enough load or storage can quickly lead to overheating the system.

Figure 7.11: *Simple Sizing Chart*

	Number of People in Household					
	1	2	3	4	5	6
	Size of Solar Storage Tank					
	40g	40g	60g	80g	90g	100g
	Square Feet of Collector					
Zone 1	20-26 ft^2	20-26	30-40	40-53	46-60	50-67
Zone 2	23-32	23-32	34-48	46-64	51-72	57-80
Zone 3	32-40	32-40	48-60	64-80	72-90	80-100
Zone 4	40-53	40-53	60-80	80-107	90-120	100-133

Additional Sizing Considerations

The methods mentioned above base the size of the system on the size of your hot water load. However, sizing the system based on your current consumption may not always be the best idea. Consider the future: a solar water heater can last more than 40 years. Do you plan to live in your home that long? You will need to consider the size of your home, how many bedrooms it has and how many people could live there. If there are only two of you now, but you live in a four-bedroom home, you may want to increase the size to accommodate a larger family, in case you sell it some day. You also need to consider the size of your own family. If you are planning on having children in the future and expect to live in the same home, you may want to size the system for additional occupants. Babies are messy; plan accordingly. However, be careful not to get carried away. A larger system is designed for a larger load, which keeps the system from overheating. Increasing the storage tank by 20–30 gallons is usually acceptable. It is important to mediate between your current situation and what you expect for the future.

You also need to take into account the clarity of your solar window and the orientation of your collectors. If your solar window has partial obstructions, but is the only option available, you will want to increase the size of your collector array by the percentage of your solar window that is obstructed. Similarly, if you are mounting at a location that is more than 30° away from due south, you will also want to increase the size of your array, probably in the range of one to two percent per degree that you are away from 30°.

Sizing a Solar Space Heating System

Let us start this section with a reality check. In most climates it is nearly impossible to heat a dwelling 100 percent with solar energy. The simple fact is that most climates have extended cloudy periods during the heating season. There are exceptions to this, such as in some areas of the northern Rocky Mountains, but most people have to deal with clouds. Your particular climate will be the single biggest influence on sizing a system. The second biggest influence will be the heat loss of your building. The reality is that systems without storage can provide an average of up to 25 percent of the annual space heating load; systems with water storage can provide an average of up to 50 percent; and high-mass systems can provide an average of 75 percent or more, even in northern climates.

This section is about heating a building. Unfortunately, all buildings lose heat to the outside when the outside temperature is colder than the inside temperature. The way a building is constructed will determine the rate at which this heat is lost. Insulation retards the flow of heat, so the more insulation a building has, the slower it will transfer the heat from inside to outside. Also, any holes in the building envelope will allow heat to escape the building and will also allow cold to easily enter. Tiny cracks around windows and plates can add up, so the tighter the building is constructed, the less air infiltration will affect the heating load. We need to consider all the above factors when calculating the heating needs of a particular building.

A well-insulated and tightly constructed building will require less energy to maintain at a constant temperature than a poorly constructed and poorly insulated building. This brings us back to something we considered in the introduction: it is cheaper to conserve energy than to purchase it. Your first task is to reduce the building's heating load as much as possible. After you have done everything that is reasonably possible to make your building more energy efficient, then you can proceed with sizing a solar heating system. Often, with existing buildings, efficiency is upgraded at the same time as adding the solar heating system.

Sizing a solar space heating system is not as straightforward as sizing a simple solar water heater. There are many more variables to consider. As we have mentioned several times in the preceding chapters, sizing solar energy systems is not a precise science because the amount of solar energy striking any given square foot of the Earth is not constant. It varies from minute to minute as well as hour to hour and year to year. The size of the system will also depend on its configuration. A dump system will require a smaller collector array than a system with storage. In addition, no two buildings are the same, so each will have its own heating characteristics. It is much easier to incorporate solar heating systems into new buildings than existing structures, where the options might be limited by how the building was constructed. We will cover these variables in the following sections and demonstrate how they will affect the size of the system.

The first step is to determine how much space heating we want. Do you want to go for the gusto, the whole banana, or would you just like to supplement with solar energy? In almost every situation, money enters the equation as a prime factor; obviously a large system requires a larger investment than a small one. In addition, some system configurations will be limited in the percentage of heat that can be delivered. Once you know your goal, you can proceed with determining the total heat load of the building for the whole year.

Calculating the Load

Whether you are installing a system with storage or a dump system without storage, the first step will be to calculate the heating load for the building. If you have an existing house that has been heated with fossil fuels, then you should have a history of the amount of fuel it takes to heat it. Your bills will show the amount of energy you used for a certain period of time.

It is impractical to size a solar heating system to heat your building on the coldest day of the year, because then for the rest of the winter your system would be oversized. A compromise must be made so your system is both cost effective and has the ability to provide a significant amount of solar heat on an annual basis. Compromise means finding a middle ground. The middle ground of the heating demand of a building is the average daily heat load of the building over the heating season. Here in central Wisconsin the average building is heated on a daily basis from October 1 through April 30, so to get our average heating load we add up the heat used for this time period and divide by the number of days over the same period. For your location, start with the month when space heat is needed on a daily basis and end when that daily heating is no longer needed. We will define this time period as our official heating season.

Let's say that you currently heat with natural gas. You get a bill each month. Let's assume there were 30 days in that billing period. Natural gas is sold by either cubic feet or therms. Let's say your bill tells you that you used 100 therms of natural gas. Each therm contains approximately 100,000 Btu. To find the average heat load of the building for this time period, take the total amount of Btu consumed and divide by 30 days. This will give you the average amount of fuel used per day.

The important figure for sizing a solar heating system is the average heat load over the official heating season, so add all the therms used during your official heating season and divide that total by the number of days in that time period. You will probably find that the average daily

Comparative Thermal Values of Fuels

Natural gas: 100,000 Btu/therm.

Coal (good stuff): 12,000 Btu/lb.

Propane (LP): 91,000 Btu/gal.

Fuel oil # 2: 139,000 Btu/gal.

Electricity: 3,412 Btu/kWh

Gasoline: 125,000 Btu/gal.

heating load for the official heating season is about 33 percent less than the average daily heating load for the coldest month of the heating season.

If you are building a new building or if you heat with electricity, you won't have the luxury of a track record to help estimate your average daily heating load. It is usually difficult to track the amount of electricity you use for heating, even if you have a kWh meter. Instead, you will have to perform a heat-load analysis of the building. An architect usually does this calculation, but many heating and solar contractors can do it as well. Heat-load calculations provided by architects and energy professionals will be expressed as Btu/heating degree day/square foot.

Most heating bills have another helpful bit of information: they display the number of heating degree days during each billing period. A heating degree day is a comparison of the outdoor temperature to 65°F. For example, if the average temperature outside for a 24-hour period was 60°F, then there would have been five heating degree days during that day. This information can be used to determine how much energy is being used to heat the building. The bill will have the total heating degree days listed for the billing period. For instance, if it was just over 30°F for every day in the month of December, you would have about 1,000 heating degree days. You just need to divide the total amount of fuel used by the total number of heating degree days. This will give you the amount of fuel used per degree day. Convert that to Btu. When you have a heat-load analysis done of your home, the calculation will result in the number of Btu per square foot per heating degree day. Now, that number alone is not sufficient to calculate your load. You need to multiply that figure times the number of square feet in your home to find out your total Btu per degree day. You will next need to find out the number of heating degree days for your geographical location. You can usually find them from your local weather service or sometimes from your utility. You will then want to tally the total for the months of your official heating season and then divide by the number of days in that season to get the average number of heating degree days per day. Multiply the result times the Btu per degree day for your entire home to get the average number of Btu per day you will need to heat your home. Presto! That's your load for solar sizing purposes.

Sizing the Collector Array

You will next want to match the size of your collector array to the size of your average load as calculated above. Because your load is based on the average number of

Btu it takes to heat your home, you will need to find out how many Btu your collectors will produce. The amount collected will depend on the outside temperature and the amount of solar radiation. Typically, flat plate panels will collect about 700–1000 Btu/ft^2/day on a clear day. In northern climates when space heating is needed, the production is usually 700–800 Btu/ft^2/day during a cold clear day in the heating season. Use this estimate, or find out the specific amount for your collector, to determine the total number of Btu/day for your collector array. For instance, a 32-square-foot panel will collect about 24,000 Btu/day.

Okay. Now we have two essential figures: 1) how many Btu are needed for an average day during the official heating season, and 2) how many Btu per square feet a flat plate collector produces each day. Divide the total number of Btu needed per average day by the calculated Btu output per day of your solar collector in your climate, and you have the number of collectors needed to heat your building on the average winter day.

Here is a sample sizing calculation. The sample home is in Wisconsin and has an average official heating season therm usage of 636 therms for the months of October through April.

This size collector array should easily provide 50 percent of the yearly heating load for the average year in central Wisconsin for the sample home. As was mentioned above, for high-mass heating systems the solar contribution will be 25 percent higher. Many locations will see greater savings than the averages for Wisconsin.

Now you have a reference point to use to start deciding what size system you want and what is practical. Undersized systems save you money. They do what they can, and every little bit helps. Large systems can significantly affect yearly heating costs. But what is the point of no return? It doesn't make sense to design for the worst day, because then the rest of the days the system will be oversized and perhaps overheating.

Another consideration we talk about is storage size and how that plays into our decision-making process.

Sample Sizing Calculation

Natural gas consumption for our official heating season:
636 therms

Days in heating season: 212 days

Average daily consumption:
636 therms ÷ 212 days = 3 therms/day

Convert to Btu: 3 × 100,000 = 300,000 Btu/day

Btu per square foot of collector: 750 Btu/ft^2

Divide total load by collector performance:
300,000 ÷ 750 = 400 ft^2 of collector

Sizing Systems with Storage

Sizing a solar heating system with storage is more complicated than any sizing we have talked about so far. The variables are collector array size, storage size, expansion tank size, pipe size and heat exchanger size. The biggest variable and the first question you have to ask is, How much of the annual heating load do I want the solar heating system to provide? You can use "rules of thumb" that have been determined over many years of experience or perform the calculations. We would suggest using both.

The discussion in the previous section concluded that for the sample house, we would need about 400 square feet of collector to provide enough Btu to heat the house for an average day. A solar storage tank sized at 1.25 gallons of storage for each square foot of collector will store enough heat for 24 hours of heating on the average day. Therefore an optimum-sized collector array would be 400 square feet, with a storage tank capacity of 500 gallons (400 × 1.25).

Rules of Thumb for Sizing Solar Heating Systems

As mentioned above, rules of thumb have been developed over the years from practical experience. They are a middle-of-the-road estimate. Your particular conditions and climate will influence these rules. The hard part of using rules of thumb is estimating the heat loss of the building that will be solar heated. These rules are designed for the average home in a moderately cold climate such as Wisconsin. In central Wisconsin we experience about 8,500 heating degree days per year. You can get the average heating degree days for your location and compare that to the central Wisconsin data. Also, if you have a highly insulated building, you must decrease the collector area. Likewise, if you live in a warmer climate, you should downsize the collector area. If you have a poorly insulated building, you must bring that building at least up to minimum standards of insulation. The big issue is overheating.

Most people's inclination is to get as large a system as is practical. They want to save the most money or have the greatest environmental impact they can. But some people back off from the biggest practical unit and decide on a smaller unit that will have some impact. It is helpful with this exercise to start with the biggest system possible and work back from there, if necessary. For sizing dump systems with no storage, we want to have one square foot of collector for every ten square feet of floor space on the main floor. We add 10 percent to the collector size for a two-story building. That sizing method should give us an annual solar contribution of 20 to 25 percent. When we add storage to a solar heating system, we can add more collector area to the array because we can store the excess heat for use later. Therefore, if we had two square feet of collector for every 10 square feet of floor area on the main floor, we should expect up to 50 percent of our annual heating load to be provided by solar energy if we have adequate storage built into the heating system.

In most cases, the 2:10 ratio will be the maximum collector array size you should consider. In a perfect situation it would be sunny every day. If this were the case, we could realistically get 100 percent of our heat from solar energy. But the reality is that we have cloudy weather. As we have mentioned, here in central Wisconsin we can experience more than five cloudy days in a row during the heating season, and we can experience as few as five sunny days in the whole month of December. No matter how many collectors we have, they do us no good when it is that cloudy. For this reason, the maximum solar contribution we can expect for a solar heating system is 50 percent unless we are using a high-mass storage configuration. Where the solar heating system is being added to an existing building, we suggest sizing the system for a 50 percent annual solar contribution. Although this is the largest size, it is a realistic goal, and will provide a reasonable return on investment.

Using the above rules, you can get in the ballpark of the collector array size. Next we need to size the storage. This part is easy. We want between one and two gallons of storage for every square foot of collector area. You should have at least 1.2 gallons of storage for each square foot of collector. If you live in a moderate climate that experiences plenty of sunny days during the heating season, use more storage.

Sizing with Computer Analysis Software

Computer programs for sizing solar heating systems with storage are available. These are convenient because they know all the right questions to ask, plus they have weather data for most locations on Earth. All the formulas needed are written into the programs. All you have to do is respond to the questions the program asks, and out comes the answer. Because these programs are fast, it is easy to vary the parameters and see how changing storage size or array size will affect the estimated yearly savings. Unfortunately, these programs are costly and are not widely distributed.

When we first started in the renewable energy business more than 25 years ago, simpler versions of these computer programs were available. Nonetheless, they were an invaluable tool to help an inexperienced installer properly size systems.

After using these programs for a while, certain patterns emerged. This is where the rules of thumb came from, as well as from watching systems over the long term. For anyone considering going into renewable energy as a profession, we suggest getting a sizing program. For the individual who is interested in putting in one or two systems, we suggest trusting the rules of thumb to size your solar energy system.

The big thing to remember when talking about sizing a solar energy system is that we are at the mercy of the weather. The sun is a variable resource. Long-term averages are a reliable indicator of system performance, but no two years are alike. Now that our weather is being influenced by the greenhouse effect, even long-term averages become less reliable.

Sizing Combination Systems

When sizing a combination domestic hot water and space heating system you must remember that with a larger system you can take advantage of economies of scale. A system experiences a certain amount of losses through the piping and related components. As the system increases in size, the losses decrease in relation to square feet of collector. For example, if your calculations say that you should have 8.5 collectors to heat the house and 2 collectors for your domestic hot water, that equals 10.5 collectors. Because the sizing you used for the domestic water portion assumed that it would be a standalone system, you can now downsize that portion of the system by as much as 25 percent because of the economies of scale. You would round down in this case to ten collectors.

Sizing Components

Sizing Pumps

Several types and sizes of pumps are used in solar water and space heating systems, but all of them fall into the general category of centrifugal pumps. In centrifugal pumps, vanes or paddles are spun within an enclosed space. Centrifugal force and the action of the paddles on the liquid create the pressure. The units of measurement used in circulating pump sizing are flow and head. Flow is the volume of liquid per minute moving through the circuit. Head is the pressure produced by a vertical column of fluid (static head) plus the pressure produced by the friction of a fluid as it flows through a pipe (friction head). Both forms of head are typically expressed in feet or in pressure.

The kind and size of pump will be determined by the needs of the particular system you are using. The biggest variety of pump sizes and types are found in the primary solar-loop pump, or main circulator. The two main types of pump used as the main circulator are high-head pumps and circulator pumps. Circulator pumps are designed to circulate a fluid within a closed loop. High-head pumps can pump fluid up an empty pipe a certain distance. Selecting the appropriate pump is critical for efficient operation of the system. An undersized pump can lead to poor flow, which may result in a diminished ability to absorb the energy from the solar radiation as the fluid passes through the collector. An oversized pump adds expense, both initially and over time. With an undersized pump, the solar fluid tends to be very hot; with an oversized pump the solar fluid does not tend to get very hot.

All systems that are completely full of solar fluid at the beginning of a heating cycle require only a circulating type of pump to operate properly. This is the case whether the closed loop is horizontal or vertical, and to virtually any height. For example, in a pressurized antifreeze system, the solar loop is full of solar fluid at all times, so when the sun comes up in the morning, the pump only has to circulate the fluid throughout the system. The head in this type of system is created only by the friction of the fluid as it circulates throughout the solar-loop piping and fittings. Circulating pumps are not designed to pump fluid up an empty pipe.

All systems that are empty of fluid in the solar loop when the system turns on require a high-head pump for the main circulator. For instance, in a drainback system the solar fluid is stored in the drainback tank when the system is off. When the system turns on, the pump must first fill the system with fluid and then continue to circulate the fluid throughout the system till it turns off, at which time the fluid then drains back into the reservoir. When filling a vertical pipe with fluid, the pump has to overcome gravity. This takes a pump with specifically designed paddles that usually operate at a higher speed. It also requires more power than a plain circulating pump. The head in this type of system is created by both the vertical pressure of the fluid in the pipe and the friction caused by the movement of the fluid within the pipes.

When sizing a pump you have to first determine whether a high-head or circulating pump is required. Then you have to calculate the head and flow requirements of the system to determine the size of the pump you will be using.

For both types of pumps, you will have to calculate the flow requirement. This is determined by adding together the flow requirement of all the individual collectors. The flow rate used for testing flat plate solar collectors by the SRCC is 0.88 gallons per minute for each collector, so that is a good flow rate to design for. However, it is always best to consult the collector manufacturer. Most manufacturers will recommend about 1 GPM, although some prefer a higher or lower flow rate. A slower flow rate keeps the fluid in the collectors for a longer period of time and results in a greater difference in temperature and a greater exiting fluid temperature, but will decrease the total amount of heat that can be collected. A faster flow rate typically results in a slightly higher efficiency, but will have a lower-temperature fluid exiting the collectors. You will need to keep in mind the design and purpose of the system. Does your application require large quantities of lower-temperature heat, as is the case for domestic hot water? Or do you need smaller quantities at a higher temperature, like what would be used for solar cooling? Because the flow rate will determine both the operating temperature and the efficiency of the system, this is an important consideration.

Figure 7.12: *Friction head chart*

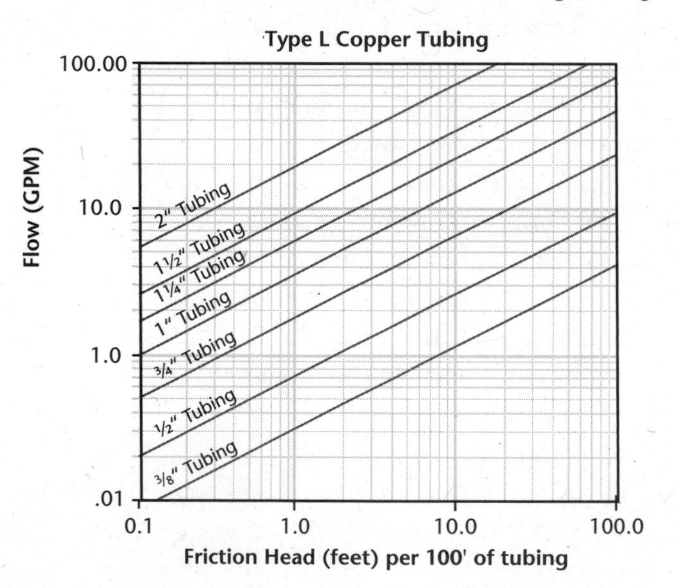

Once you have decided on a flow rate, you will next want to determine the total amount of head pressure. Friction head can be calculated by adding the standard amount of head for each foot of pipe in the circuit plus the head pressure for each fitting and component in the circuit. Figure 7.12 illustrates the amount of friction head per 100' feet of pipe at different flow rates and pipe sizes for Type L copper. As the flow rate increases, the head increases. Similarly, as the pipe size increases, the head decreases. The friction head for fittings and valves is typically provided in "equivalent length of tubing." For example, a ¾" Ell will have the same amount of friction head as 2' of straight tubing. Figure 7.13 lists some of the more common items found along the solar loop. Some plumbing components, such as heat exchangers or collectors, are individually rated for head pressure. This information is often supplied in the product specifications. The friction head for some collectors can often be found on the SRCC data sheet, although it is oftentimes provided in pounds per square inch (PSI) instead of feet, so a conversion will be required. 1 PSI = 2.31 ft. of head, or 1 ft. of head = .4335 PSI.

Head pressure created by the pressure of the vertical height (static head) of the

Pressure Loss in Sweat Fittings and Valves (Equivalent Length of Feet of Tubing)									
	Fittings						Valves		
Nominal Size	90° Ell	45° Ell	90° Tee	Straight Tee	Coupling	Ball	Gate	Butterfly	Check
3/8	0.5	0	1.5	0	0	0	0	0	1.5
½	1	0.5	2	0	0	0	0	0	2
5/8	1.5	0.5	2	0	0	0	0	0	2.5
¾	2	0.5	3	0	0	0	0	0	3
1	2.5	1	4.5	0	0	0.5	0	0	4.5
1¼	3	1	5.5	0.5	0.5	0.5	0	0	5.5
1½	4	1.5	7	0.5	0.5	0.5	0	0	6.5
2	5.5	2	9	0.5	0.5	0.5	0.5	7.5	9
2½	7	2.5	12	0.5	0.5	0	1	10	11.5
3	9	3.5	15	1	1	0	1.5	15.5	14.5

COPPER TUBE HANDBOOK

Figure 7.13: *Pressure loss in fittings and valves*

circuit will next have to be calculated. This will be the case only on drainback systems. The static head pressure is simply the linear vertical distance from highest point in the system, typically the top of the collector array, to resting fluid location, typically the top of the drainback tank. Note that it is not distance from the pump to the collectors, because the pump will always be below the water level when it has drained back into the tank. In order to minimize static head, the drainback tank is best located as close to the collector array as possible. As a safety factor, you may want to add an additional three to four feet of static head on drainback systems. We have seen several cases in which the drainback pump was undersized. The pumps were just large enough to lift some fluid into the collector but too small to completely fill it up and complete the circuit. The small amount of fluid that made it up into the collector would overheat and boil.

Once you have calculated the total head pressure and flow rate of the circuit, you can then select the perfect pump for your job. All pumps have a pump curve, which is a graph that shows how the pump will perform at various flow rates and head pressures. These pump curves are supplied by pump manufacturers, or you can find them on the Internet. Just type the brand and model of a pump you are considering into your search engine and you will find it. You can also probably jump from there to a manufacturer's main page and find other pumps they offer and see their pump curves as well. Wholesale suppliers also provide pump curves.

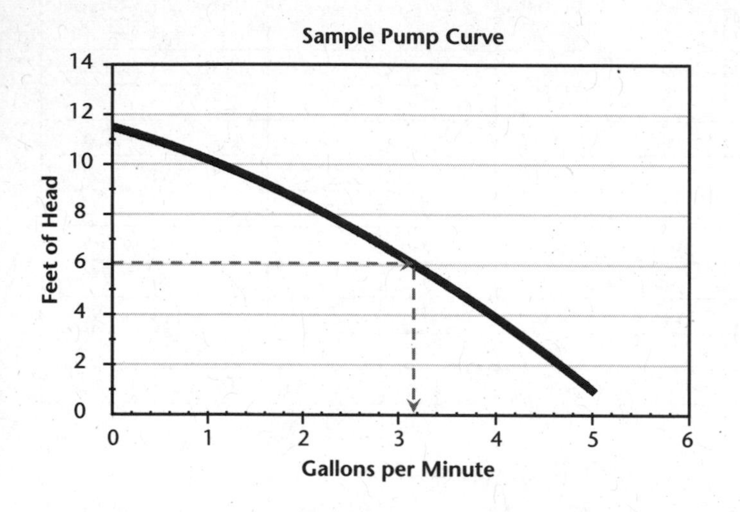

Figure 7.14: *Sample pump curve*

The sample pump sizing calculation demonstrates the steps outlined above. In this case, the pump will supply slightly more flow, 3.15 gpm, than for what the system has been designed for, 3 gpm. If you get this close in real life, this would be a great pump choice. You can always add a little extra head by cracking a ball valve on the loop and slowing the pump down if it is slightly too big, but you can never reduce the head in the system to get more flow.

The pump curves provided by manufacturers represent the pumps' performance when they are pumping water. Propylene glycol is slippery and is harder to pump

Sample Pump Sizing Calculation

System type	Pressurized antifreeze
Number of collectors	3
Friction head for collector array	.75 feet of head
Total system design flow	3 gpm (1gpm/collector)
Tubing size	¾″ Type L copper
Linear feet of tubing	100 feet
Equivalent feet of tubing due to fittings	33 feet equivalent
Friction head for tubing and fittings	4 feet of head (¾″ tubing @ 3 gpm = 3 feet of head per 100′ of tubing)
Friction head of heat exchanger	1.25 feet of head
Total friction head	6 feet of head
Total static head	0 feet of head
Total head pressure	6 feet of head
Flow for sample pump at total head pressure	3.15 gpm

than water at lower temperatures, but once it gets hot it is comparable to water. When pumping at cooler temperatures, a correction factor may need to be applied.

When installing pumps you need to ensure that there is a sufficient length of straight tubing both before and after the pump. The generally accepted rule is that there needs to be four times the diameter of the tube size on the suction side and 2.5 times the diameter on the discharge side. For instance, if you are using 1" copper tube, you will want at least 4" of straight pipe before the pump before you insert any fittings. If this rule is not followed, you may cause restriction on the pump and generate unnecessary head pressure.

Pipe sizing

The size of the piping to be used for the solar loop is typically determined by the amount of flow that is required. Piping that is too small will cause excessive friction, reduce flow and require a significantly larger pump. Piping that is too large may result in an increased and unnecessary cost for both the piping and solar fluid. Generally, you will want to use ¾" piping for four or fewer collectors and 1" for up to eight collectors. For other sizes of arrays or for split arrays, consult Figure 7.15 for the recommended maximum flow for common pipe sizes. You can always force a higher flow through a pipe than what is recommended, but this will result in having to purchase a larger pump, which will consume more electricity. The increase in operating costs in the long run will offset any savings from using smaller pipe. It is always preferable to have too large a pipe than too small, especially if you ever plan to expand the size of your system.

Pipe Size	Recommended Maximum Flow
½"	1½ gpm
¾"	4 gpm
1"	8 gpm
1¼"	14 gpm
1½"	22 gpm
2"	45 gpm
2½"	85 gpm
3"	130 gpm

Figure 7.15: *Recommended maximum pipe sizes*

Fluid Volume Estimation

The amount of solar fluid in a system will be a factor of the size of piping, the length of piping and the fluid requirements of the collectors, heat exchanger and any other components. In a pressurized system the solar loop will be completely full of fluid, so you will need to determine the total volume capacity of the whole loop. First, find out how much fluid the collectors will hold. Flat plate collectors will hold anywhere from 1.25–2 gallons of fluid. Evacuated tube collectors typically hold much less. In either case, check with the manufacturer or SRCC to get an accurate amount. Next, determine how much fluid will be required for all of the piping. Figure 7.16 demonstrates the amount of fluid necessary to fill the various types of tubing commonly used in the solar loop. Multiply the number provided by the total linear length of tubing in the system. Finally, add in the amount of fluid used in any additional components in the system. Most components hold a negligible amount of fluid and can be ignored, but some heat exchangers can make quite a significant difference. For instance, indirect tanks that are the jacket style can hold up to as much as six gallons of solar fluid. In a drainback system, part of the solar loop will be empty when the system is not functioning. The fluid requirement will be the sum of the drainback tank and all of the piping below it.

Sizing Expansion Tanks

An appropriate-sized expansion tank is critical on all pressurized systems. If the tank is undersized, you will have dramatic swings in pressure. If the pressure gets too high, you may set off the pressure-relief valve or possibly burst one of the joints in the system. If the pressure gets too low when the system cools down, you may suck air into the system. Every expansion tank manufacturer will have guidelines for sizing its particular product based on the total amount of fluid in the system and the anticipated temperature swings. Because solar water heating systems experience much greater swings in temperature than typical hydronic systems, you will want to carefully follow the manufacturer's recommendations. Figure 7.17 demonstrates some basic guidelines for sizing for some common nominal expansion tanks.

Figure 7.16: *Fluid volume estimation*

Gallons of Fluid per Foot of Tubing				
Nominal Size	Type K Copper	Type L Copper	Type M Copper	hePEX
3/8"	0.00660	0.00755	0.00827	0.00500
½"	0.01134	0.01212	0.01321	0.00921
5/8"	0.01735	0.01810	N/A	0.01345
¾"	0.02265	0.02515	0.02685	0.01838
1"	0.04041	0.04288	0.04543	0.03033
1¼"	0.06327	0.06532	0.06803	0.04668
1½"	0.08952	0.09245	0.09517	0.06516
2"	0.15664	0.16083	0.16474	0.11160
2½"	0.24201	0.24801	0.25408	N/A
3"	0.34492	0.35400	0.36271	N/A

Sizing Air Heating Systems

System Fluid Volume	Expansion Tanks
4.7	1 - #15
9.4	2 - #15
12.5	1 - #30
17.2	1 - #15, 1 - #30
25	2 - #30
44.5	1 - #90
56	1 - #30, 1 - #90
88	2 - #90

Figure 7.17: *Expansion tank sizing*

The important consideration when sizing air systems is to not oversize the collector. A balance must be met among collector output, building heat loss and amount of thermal mass within the structure. In cold climates, the rule of thumb for sizing air systems is that ten percent of the footprint of the building equals the collector size. This is for an average building. For buildings with high thermal performance, the collector size should be reduced appropriately. For buildings with large amounts of thermal mass or poor thermal performance, the collector size can be increased accordingly. As a double check, it is helpful to compare the heat loss of the building to the rated collector output. Remember that the collector will be operating about four hours per day during the winter, so it has to do all its work in a relatively short time frame. In buildings with large amounts of thermal mass, collector output can exceed building heat loss by up to 100 percent. Defer to the collector manufacturer for help regarding sizing, especially when using transpired collectors. However, you will be doing calculations similar to those above for liquid-type systems by calculating the average heating load and collector output and matching the two.

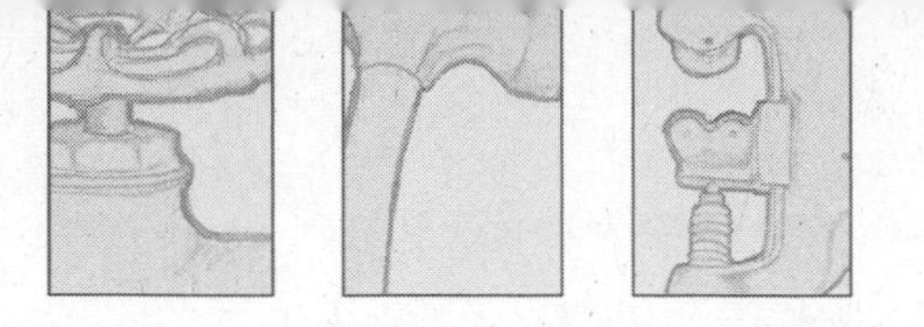

8

SYSTEM INSTALLATION

IN THE PREVIOUS CHAPTERS we described most of the technologies used today for solar water heating. The rest of the book covers details about installing and operating solar water and space heating systems. Most of the systems we have talked about share many principles. As you have seen, most systems are variations on a similar theme. They therefore share many components and installation requirements. We will detail information about specific technologies as we work through the basic installation principles.

Installing Solar Water Heating Systems

A characteristic of every successful installation is proper and thorough planning. An installer with solid understanding of how a solar water heating system operates and what basic components are used for a particular system can visualize how that system will be incorporated into a structure. There are almost always several optional configurations to be considered. The options must be weighed against each other to determine the best of all possible locations for each component. The load analysis, which is the first task in the process, will determine how many collectors will be used. The solar window will determine the best location for the collectors. The location of the collectors and the layout and type of building will dictate where the main solar-loop piping will be located. The area where the existing water heater is located will most often identify the area where the heat exchanger and the solar storage tank will be located. You need to visualize the system with all its parts already installed on the building.

Once you have decided where all the components will be located, it is time to make a list of all the necessary parts. A good method of making sure you have them all is to start at the top and work your way down. It is also helpful to make a sketch of the system, which will help identify the needed components. You have already determined how many collectors will be used. Next, what type of mounting hardware will be used for the collectors? For roof-mounted collectors, determine how the piping will penetrate the roof, and identify the proper roof flashing. Measure the pipe runs next, from the collectors all the way to the heat exchanger and storage tank, and be sure to add the appropriate pipe insulation. Now go back and identify all the pipe fittings that will be needed, including elbows, tees, unions, adapters and pipe hangers. While you are doing this, visualize how you might actually get the pipe installed in the areas you have identified. Is there enough room to maneuver each pipe into its final location, or will you have to cut short lengths to accommodate the installation? Make sure you have plenty of couplers to join straight lengths of pipe together. The size of the collector array will determine what size storage tank will be required and what size heat exchanger or heat exchange module you will need. If you are not purchasing a preplumbed heat exchanger module, you will need to identify all the components required to build your own. These parts may include boiler drains, check valves, drainback tank, thermometers, pressure gauge, pressure-relief valve and others.

Once you have all the piping figured out, you can calculate how much solar fluid will be required. Fluid volume estimation is covered in the previous chapters. Please refer to the system schematics in earlier chapters to make sure you have everything you will need. Also remember never to mix piping of dissimilar metals when installing a solar water or space heating system — or any plumbing system, for that matter. Compatible metals include copper, stainless steel, brass and bronze. Noncompatible metals are cast iron, galvanized steel, black iron and aluminum. (Actually, all iron and steel pipes are compatible among themselves, but are not compatible with the other types of metals mentioned.) Whenever noncompatible metals must connect, as where the copper piping attaches to a steel storage tank, use a dielectric union at each connection point.

Professional installers and plumbers usually carry a large assortment of parts with them to every job site. Bob used to call his installation truck his rolling hardware store. Even if you have your own rolling hardware store, it is important to check your inventory before you leave for the work site to make sure you have everything. Having to run to the plumbing supply house several times will add frustration and additional costs to any installation. We recommend having extra parts on the work site, especially pipe fittings, soldering supplies, solar fluid and consumables such as fasteners. This is especially critical for some of the proprietary fittings or mounting hardware components that are specific to your collector or system. These are typically things that cannot be purchased at a local hardware store, so make sure you have a sufficient supply.

Some states or municipalities require professional installers to be licensed. Be sure to check with local authorities to find out if any licenses are required. Building permits may also be required, so check with the local building inspector.

The last part of the planning process is to make sure you have all the right tools and supplies to facilitate the installation. On page 162 is a list of typical installation tools and supplies. A good solar installer will have all of them handy in his or her toolbox.

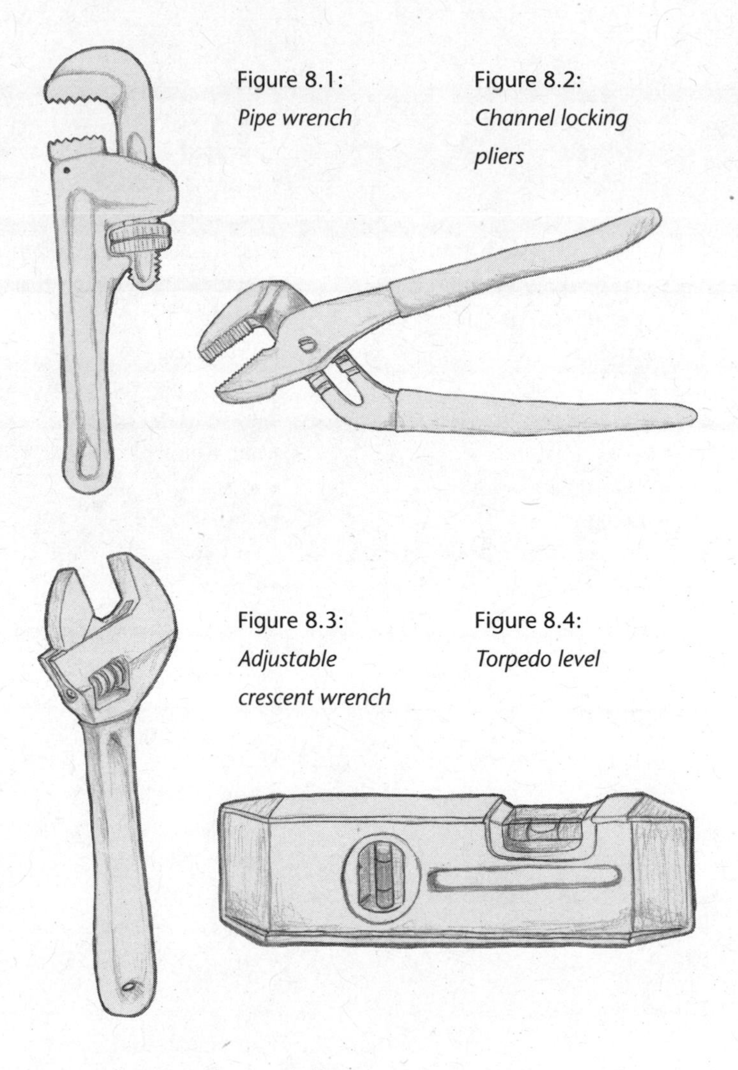

Figure 8.1: *Pipe wrench*

Figure 8.2: *Channel locking pliers*

Figure 8.3: *Adjustable crescent wrench*

Figure 8.4: *Torpedo level*

Handling Solar Collectors

Shipping

As we said before, most installations will start with the collectors. However, first you need to get them. Collectors are typically shipped stacked on a wooden pallet. When you receive your shipment of collectors, carefully inspect each collector for shipping damage. An obvious indication of damage will be broken glass. Other damage could be dents to the collector

Tools

- soldering torch with extra fuel
- pipe cutter
- deburring tool
- fitting brushes
- soldering paste brush(es)
- charging pump and hoses
- pressure gauge (automotive type)
- tape measure
- marking pens and pencils
- hacksaw
- assorted screwdrivers
- assorted pliers
- electric drill(s), including battery-operated type
- assorted drill bits, including assorted hole saws
- hammer(s)
- assorted pipe wrenches
- channel locking pliers
- assorted crescent wrenches
- assorted socket wrenches
- reciprocating saw with assorted blades
- putty knife
- utility knife with extra blades
- level
- angle finder
- square
- wire stripper
- tin snips
- crowbar
- stepstool or short folding stepladder
- six- to seven-foot folding stepladder
- extension ladder(s)
- roof jacks with planks
- safety harnesses and safety equipment
- rope
- first aid kit
- five-gallon pails
- work gloves
- safety glasses
- flashlight and assorted work lights
- extension cords
- sawhorses
- camera
- crane or lift (optional)

Supplies

- lead-free solder
- soldering paste
- grit cloth
- pipe dope
- Teflon tape
- assorted fasteners
- screws
- nails (including roofing nails)
- stainless nuts, bolts, washers
- roof cement
- caulk
- assorted zip ties
- electrical tape
- insulation adhesive
- identification labels
- rags
- sensor wire or electrical wire
- wire nuts or electrical solder

frame or back. Any damage should be written on the bill of lading. If the damage is severe, you may choose not to accept the shipment. If the damage is minor, you can accept the shipment and negotiate a settlement with the shipper for repair costs. As long as the collectors are structurally sound, they will probably be okay.

You usually have two options for receiving the shipment. You can have the collectors shipped directly to your site, or you can pick them up at the local trucking terminal. If you have them delivered, you will have to unload them off the back of a semitrailer. This is easy if you have a loading dock and even easier if you have a forklift. If you don't have access to a loading dock, the next best thing is to back a truck up to the tailgate of the semitrailer. If you are receiving more than one or two collectors, it will probably be easiest to open the crate while it is still on the semi and then unload the collectors individually. Most semi drivers will help you unload, but we suggest having two people available when delivery is expected. Also have a tin snips and crowbar ready to open the crate. Remember that the drivers have schedules to keep and usually don't have all day to fool around unloading one shipment. When you order your collectors, you can request a phone call from the trucking company alerting you to the exact delivery time. There may be a small fee for this service, but it is usually worth it because you can be prepared to accept your delivery and have your help there.

The other option is to have your collectors kept at the local trucking company terminal until you can pick them up. There are several advantages to this option. First, shipping is cheaper if the trucking company does not have to deliver right to your place. Second, you don't have to wait around for your delivery to arrive. You can go to the terminal at your convenience and take your time uncrating the collectors. You will need access to a truck or trailer that is large enough to haul the collectors to your site. Note that most trucking firms will start to charge you a storage fee if you do not pick up the collectors within a reasonable time, usually one week.

Each collector weights 85 to 140 pounds. If the collectors will be transported a short distance, they can be laid flat, with the glass side up. If they need to be stacked, place cardboard or two-by-fours between the collectors, taking special care not to let them slide around on each other. Collectors can be easily damaged, so before transportation securely tie them down in such a way that they will not shift.

On one hand, collectors are very tough pieces of equipment that will last decades on the roof of your dwelling. On the other hand, during handling and transportation, the collector manifolds and glazing can be easily damaged, and the glazing strips can be easily scratched. To prevent damage, always handle the collectors only by the frame, never by the manifold pipes protruding from the frame. If you pick up the collectors by these manifolds, they can bend and get out of round, causing problems later in the installation process. If the collectors have been sitting in the sun, they can get very hot, especially the manifolds. Always use gloves.

The best place to store collectors prior to mounting is inside a building, out of the weather. Collectors can be stored outside if well covered. Never leave the collectors exposed to the direct sun for any period of time because they will get very hot and can cause burns if touched with bare hands. This is especially true with evacuated tube collectors. Don't cover them with clear plastic because they will get too hot.

Most collectors are shipped with plastic caps on the manifold ends. These are intended to keep moisture and dirt from entering the collectors. If the collector is exposed to prolonged sunlight, these manifolds can get very hot, as mentioned above. The plastic caps can either blow off or melt onto the manifolds, causing additional problems during final installation — all the more reason to keep the collectors well covered. Some manufacturers also tape a plastic bag to the glass, which contains installation instructions and warranty cards. Remove this as soon as possible so it won't get lost or melted to the collector.

Getting the Panels onto the Roof

Carefully plan how you are going to get the collectors onto the roof. Always remember: safety first. Don't try lifting the collectors when it is very windy or gusty. Always keep bystanders, especially children, away from the action. Whenever possible, use scaffolding, roof jacks and planks when working on a roof. Always use safety ropes and approved fall-protection harness. Be sure to adhere to any OSHA safety guidelines. In most cases it is required to wear a safety harness when the work area is more than six feet high.

Sneakers or running shoes give a better grip than leather-soled shoes. Rubbers also give a good grip. Heavy lugged boots tend to tear up shingles, especially when the roof is hot. Avoid walking repeatedly on the same area of the roof to keep wear to a minimum. The use of planks or plywood on top of the shingles also saves roofing from wear. Don't walk on the ridge. The steeper the roof, the higher the danger. Make sure your safety ropes are securely tied to something that will hold when necessary. Take your time, and don't rush.

The easiest and safest way to get collectors onto the roof is to use some mechanical lift, such as a bucket truck, crane, forklift or roof hoist. Many roofers have devices to place shingles on roofs, and these can work well. Some companies now sell portable ladder lifts that have a ledge specifically for lifting solar panels onto roofs.

Lifting collectors to the roof manually is also possible. We suggest having at least three people for this procedure. Set up two ladders next to each other and leading to the roof. Have one person on the roof and two on the ground. Install the mounting brackets onto each collector. Lean a collector on the ladders with one end on the ground. Use a rope long enough to reach from the roof to the ground, and tie the lower end to the collector. Tie the rope to the mounting brackets — never to the manifold pipes. When ready, the workers on the ground should slide the collector up the ladders while the person on the roof pulls on the rope. As the collector slides up the ladders, the two workers can walk it up till the collector reaches the top of the ladders.

At this point, the person on the roof needs to set the rope aside and grab the top of the collector. The collector then can continue to move up until half of it is above the top of the ladders. Now pivot the top of the collector down toward the roof. You must be careful at this point not to rest the back of the collector on the top of the ladders, as this will dent it. Hold the collector just above the top of the ladders. The worker on the roof can move up the roof with the top of the collector until the bottom of the collector passes the ladders. The two workers on the ladders must hold the collector securely throughout this whole process. After the collector is lying on the roof, the workers on the ladders can climb onto the roof and help the other worker move the panel to its final location. If the roof is steep, have roof jacks and a plank set up near where the bottom of the collectors will be mounted. The roof jack and plank will keep the collector from sliding off the roof.

Mounting Solar Collectors

A previous section was devoted to siting your solar collectors. This process identified potential locations for your collector array. After considering all the variables, you have probably chosen to mount the collector array either on a roof or on a rack on the ground. We will address each of these mounting methods. Much of what follows describes the mounting of flat plate solar thermal collectors, but mounting evacuated tube collectors is principally the same because the mechanical integration methods are often identical.

Collector Tilt

Before you begin to mount your collectors, you will need to know their collector tilt — the angle the surface of the collector makes with the horizon. As we have mentioned, solar collectors will perform at optimum efficiency and maximum output if directly facing the sun. We have observed how the sun tracks across the sky from morning to night. Wouldn't it be nice to have our collectors face east in the morning to collect the early morning sun, then follow the sun's path all day long till sunset? A good idea, but unfortunately our collectors need piping to bring the heat they collect into the house. There are no flexible piping products available today that would allow them to follow the sun. Furthermore, because of the weight of the panels, the amount of energy necessary to move the panels would nearly offset any gains. We therefore have to utilize a fixed array. Other solar technologies, such as photovoltaic arrays, can track the sun's path across the sky because they deliver electrons through wires, which are flexible, and the panels are typically much lighter.

We have noted that the angle of the sun at noon will vary depending on the season of the year, being lower in the winter and higher in the summer. Because we cannot seasonally adjust the angle at which our collectors are mounted in order to maximize collector performance, we need to choose a permanent mounting angle that will be best for year-round performance. This mounting angle will be a compromise between summer and winter optimums.

Collector tilt will be determined by several factors. The following sidebar describes basic rules for determining collector tilt.

For a solar hot water system that will be used year round, a tilt angle equal to the site's latitude is considered best. You can get your latitude from a map or from your local weather bureau office. For a solar water heater that will be used only during the summer, subtract ten degrees from your latitude to get your optimum tilt angle. Because solar heating systems are intended to work best during the winter months when the sun is low in the sky, the collectors must be tilted to maximize this low sun angle. Mount your collectors at an angle that equals your latitude plus 15 degrees. Here in Amherst, Wisconsin, our latitude is 45 degrees, so we mount our collectors at a 60-degree angle. People often comment that if their system is a combination space heating and water heating system this high angle might decrease its efficiency during the summer when they still need hot water. This is true, but there will be many collectors in this system, many more than are required to heat the domestic hot water. So the bottom line is that you will still have more hot water than you could possibly use with the collectors mounted at this tilt. If you plan on a year-round system for just domestic hot water and you live in an area that gets significant snow, your collectors should be mounted at an angle of more than 40 degrees to help the snow slide off the collectors.

Collector Mounting

Solar thermal collectors are a lot like big sails. Because of this, they need to be securely mounted. The mounting system needs to withstand winds of 100 miles per hour and gusts up to 150 miles per hour. In coastal areas where hurricane force winds are possible, extra precautions should be observed when mounting collectors. The wind pressure on a solar collector, especially when blowing from the north, will create pressure in every direction, including lift. This lifting force of the collector is very strong. Plan accordingly. We have seen people install collectors using screw-type fasteners and have their collectors blown off their roofs. It was not a pretty sight.

Collectors' mounting systems can be adapted to varied situations. Most manufacturers of thermal collectors also make mounting hardware for their collectors. Mounting brackets, usually four per collector, are attached to the collector frame. These brackets can be bolted directly to a roof, bolted to extension legs or bolted to a rack. Every manufacturer has a unique way of attaching a bracket to its particular collector, so you want to purchase the brackets from your collector manufacturer.

The brackets usually can be located anywhere along the collector frame, and can be tightened securely in many positions. Often, the bracket is shaped like an L. Brackets can be located on the top and bottom or on both sides of the collector. Usually, brackets bought from the manufacturer can be used for parallel roof mounting and also for rack mounting. Sometimes there are special brackets used for attachment to adjustable legs.

If your collectors do not have mounting hardware included, and the manufacturer is no longer in business, you will have to come up with your own method of attaching the collectors to the roof or rack. Some collectors have a lip or flange running around the perimeter of the collector, and you can often drill and bolt directly to that flange. If no flange is available, you will have to attach homemade brackets to the sides of the collector. When attaching the bracket to the side of the collector, it is important not to drill a hole into the absorber plate or any piping within the collector. Look carefully through the glass to see where these are and drill to miss them. We typically use aluminum L-brackets in this situation. Be sure to use stainless tech screws to attach the L-bracket to the collector. The sides of a collector are the best place to mount these L-brackets.

Flat plate collector frames and mounting hardware are typically made of aluminum. Because galvanized bolts and washers rust quickly when they contact aluminum, always use stainless steel for all

System Use
(Example @ 45 degrees latitude)

Domestic hot water, year-round usage, tilt angle: Latitude 45 degrees

Domestic hot water, summer use only, tilt angle: Latitude –10 degrees = 35 degrees

Space heating/domestic hot water combination, tilt angle: Latitude +15 degrees = 60 degrees

fasteners. If stainless fasteners are not available, use a neoprene or rubber washer between all dissimilar metals.

Many flat plate collectors have weep holes on one end. Most collectors are rectangular, and the weep holes will be on one of the short ends. Mount the collectors with the weep holes on the bottom. These weep holes allow moisture to escape from the collector and allow the collector to maintain atmospheric pressure. Otherwise, they could build up pressure when they get hot, which could pose problems.

Roof Mounting

When identifying the final location on the roof for the collectors, be sure to notice any potential shadows that could be cast by chimneys or other sections of the roof onto the collectors, and adjust the array location accordingly. Also look for shading from trees or other buildings. It is usually best to have the top of the array as far up the roof as possible. This will allow easier fastening of the collectors to the roof structure and will also reduce the amount of snow that could accumulate behind the collectors.

It is important to locate all power lines in the working area. It is dangerous to be handling collectors near electrical lines. You risk electrocution if solar collectors come in contact with electrical lines. Avoid positioning collectors where there is a possibility of contact. If the very best array location has power lines nearby, consider having the power lines moved before attempting to mount the collectors.

Make sure that the roofing materials are in good shape before mounting the collectors. If the shingles need to be replaced, it is wise to do this before mounting the collectors. Wood shake and tile roofs are very fragile, and extreme care must be taken when walking and mounting collectors on them. Remember to use planks and plywood to protect roofing. Most installers like to get as much of the roof work as possible done early in the day, as it is often cooler then. Also avoid walking on a hot roof, because when shingles get hot they are soft and more susceptible to damage. It is typical to install a set of temporary roof jacks and a plank just below where the bottom of the collectors will be located. The roof jack and plank arrangement makes a stable place to stand as well as providing a safety barrier to keep collectors from sliding off the roof.

Flat plate and evacuated tube collectors weigh only about three to four pounds per square foot, so they do not pose a significant weight problem for most roofs. Nonetheless, it is always good to inspect the roof structure to make sure there are no rotten roof members. Someone will be walking around up there, and you don't want anyone falling through the roof.

ICS collectors can pose a significant weight problem because they can hold a large amount of water inside the collectors. Careful calculations should be made to guarantee that the roofing system will hold the additional weight. Additional roof bracing may be required. If you are not qualified to do these calculations, please consult an architect or engineer for advice.

Hardware Mounting Methods

As mentioned above, the collector array must be securely attached to the roof in order to withstand wind speeds of 100 miles per hour or more. The three most popular methods of attaching mounting hardware to roofs are (A) lag bolts or regular bolts secured through the roofing materials and into spanners installed under the sheathing and between the rafters, (B) lag bolts secured through the roofing materials and into the rafters and (C) long bolts secured through the roofing materials and through spanners located on the underside of the rafters.

Of the three options mentioned above, C is the strongest and is the method recommended for all installations, if possible, especially if high winds are experienced at your location. Options A and B are also very strong and can be used in most locations and for attaching extension legs.

Option A shows a roof bracket that is bolted to a spanner located under the sheathing. The spanner should fit tightly between the rafters. The spanner can be a two-by-four or two-by-six. The spanner should be pushed tight to the sheathing and secured to the rafters with two nails or screws on each end of the spanner. Predrill your bolt holes slightly larger than the bolt size and use a large washer on the end of the bolt. This method works well if there is to be a finished ceiling attached to the rafters.

Option B shows a roof bracket that is lag bolted into a rafter. Always predrill holes for the lag bolts. If lag bolts are screwed into rafters without predrilling, there is a good chance you will crack the rafter, resulting in a weak fastening. Use a drill bit that is smaller than the lag bolt so there will be wood left for the threads to bite into, and drill the hole slightly

Figure 8.5: *Hardware mounting methods*

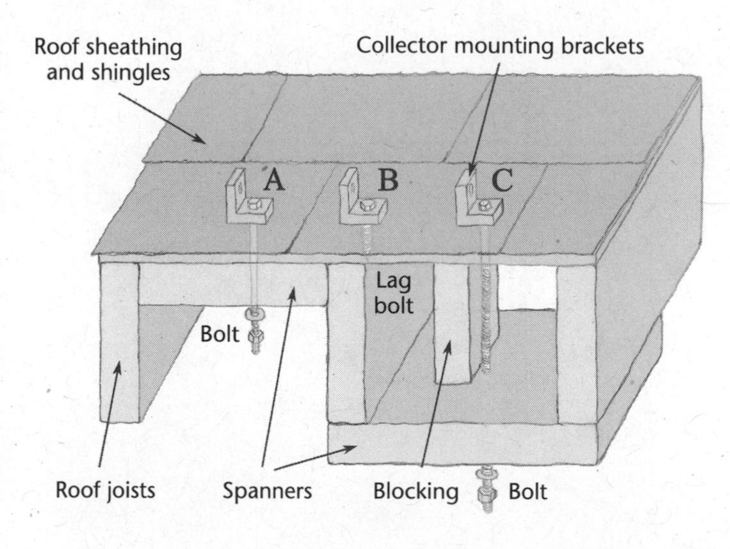

shallower than the length of the bolt so the end of the bolt screws into solid wood. If you are using a spacer between the mounting bracket and the roof, drill a hole completely through the spacer, using a drill size slightly larger than the largest diameter of the bolt so it will slide through the spacer easily.

Option C shows a roof bracket that is bolted to a spanner located on the underside of the rafters. A long bolt or threaded rod is used to fasten the bracket to the spanner. Some installers use a piece of Unistrut here as the spanner.

Wood spacers that are the same dimension as the rafters are used between the sheathing and the spanner to prevent the roof sheathing from caving in when the nut is tightened.

Figure 8.6: *Determining roof pitch*

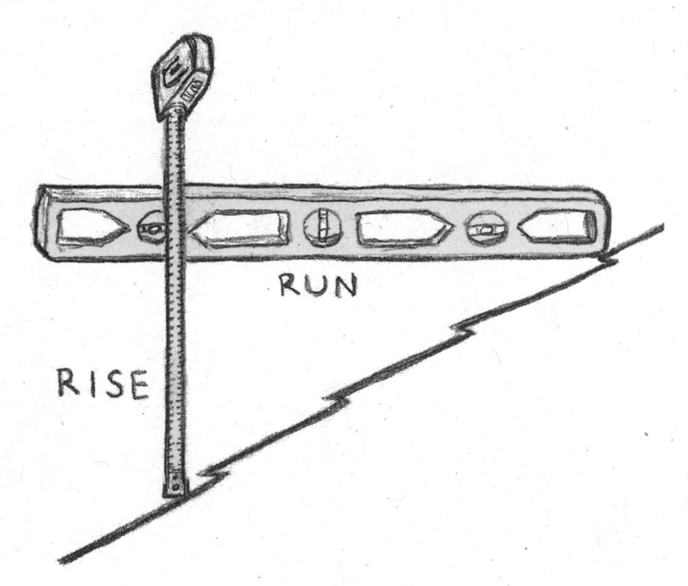

For flat roofs, a rack needs to be built to hold the collector array at the proper angle and orientation. The rack will be similar to ground-mount racks. For flat roofs, it is best to contact the installing roofer or a roofing professional. The roofer will install a curb, or pitch pot, which is a waterproof pier to which you can attach the roof brackets. Remember to consider the strength of the mounts, which have to hold the collectors down during high wind conditions. Remember, too, that the panels want to lift up during high winds, so plan accordingly.

Whenever a hole is put into the roof, adequate sealant must be used to assure a leak-proof installation. Use liberal amounts of butyl sealant or roofing cement inside the roof penetrations, under mounting brackets and shims, and around all fasteners.

Roof Pitch

When mounting collectors on a roof, you must determine the pitch of your roof. Roof pitch describes the number of inches a roof drops per foot. You can determine your pitch by using a level and a tape measure. (See Figure 8.6.) Place one end of a level on the roof, hold it perpendicular to the roof and level horizontally. Measure 12 inches out from the roof on the level, and that will be your run. Now measure the distance vertically from the 12-inch mark on the level down to the roof, and that will be your rise. For instance, you might measure four inches down. This equates to a 4:12 pitch or approximately 18°.

Use the chart in Figure 8.7 to get your tilt angle from the roof pitch. Solar professionals may want to purchase an angle protractor, also called an angle finder, for a quick way to find the angle. To use an angle protractor, you simply place it on the roof and the needle will swing to the appropriate angle.

Roof Pitch	Tilt Angle(degrees)
0:12	0
1:12	5
2:12	9
3:12	14
4:12	18
5:12	23
6:12	27
7:12	30
8:12	34
9:12	37
10:12	40
11:12	43
12:12	45
14:12	49
16:12	53
18:12	56
20:12	60

Figure 8.7: *Converting roof pitch to degrees chart*

Riser Legs

If the roof pitch equals the angle you have determined is optimal for your area, you can mount the collectors flush to the roof. In this case, the collectors will be held two or three inches above the roof, using mounting brackets and spacers (if needed). If the roof pitch is less than the optimum mounting angle, the collectors will need to be tilted. This is accomplished by raising the top of the collectors with riser legs that are attached to the roof with mounting brackets. The riser legs will need to extend at a 90-degree angle from the roof. The length of the riser legs will vary with each installation according to the pitch of the roof, the collector mounting angle and the length of the collector. Below is a chart you can use to determine the length of the riser legs for the collector. This particular chart is intended for 48"-wide collectors mounted horizontally. If you plan to mount a 4' × 8' collector vertically, you will need to double the length. For a horizontal 4' × 10', multiply the figure by 2.5. Negative numbers in the chart mean that you need riser legs in the

Figure 8.8: Angle protractor

front of the collector instead of the back. For odd-shaped collectors, you can use the following trigonometric calculation to determine the length:

$$S = 2L \sin \frac{T-P}{2}$$

Where:

S = Standoff Length,

L = Vertical Length of Collector,

T = Tilt Angle of Collector,

P = Pitch Angle of Roof

Installing Roof Mounts

Before attempting to hoist the collectors to the roof, it is important to plan how you are going to attach them and to have all the necessary components assembled. You will need access to the underside of the roof during the collector mounting process, so have this access ready beforehand as well. Likewise, have all your tools gathered before starting.

Once you have identified a potential roof area for collector location, it is time to lay out a plan on the roof to make sure that you have enough room for the array. Measure your collectors or look on the spec sheet for the collector to get the exact collector size. We talk about generic collector sizes such as 4' × 8' or 4' × 10',

Figure 8.9: *Determining riser leg length*

Roof Slope		Collector Tilt							
Pitch	Angle	30 degrees	35	40	45	50	55	60	65
Flat	0 degrees	25"	29	33	37	41	44	48	52
1:12	5	21	25	29	33	37	41	44	48
2:12	9	17	22	26	30	34	38	41	45
3:12	14	13	17	22	26	30	34	38	41
4:12	18	10	14	18	22	26	30	34	38
5:12	23	6	10	14	18	22	26	30	34
6:12	27	3	7	11	15	19	23	27	31
7:12	30	0	4	8	13	17	21	25	29
8:12	34	-3	1	5	9	13	17	22	26
9:12	37	-6	-2	3	7	11	15	19	23
10:12	40	-8	-4	0	4	8	12	17	21
11:12	43	-11	-7	-3	2	6	10	14	18
12:12	45	-13	-8	-4	0	4	8	13	17
14:12	49	-16	-12	-8	-3	-1	5	9	13
16:12	53	-19	-15	-11	-7	-3	2	6	10
Vertical	90	-48	-44	-41	-37	-33	-29	-25	-21
		length in inches							

but actual sizes vary for every model. Collectors are typically slightly larger than the generic size. Also calculate the additional distance that will be required between collectors.

Most collectors have about 1.5 inches of manifold pipe protruding out their sides near the top and bottom. These pipes facilitate plumbing the collectors together. Some collectors have factory-installed unions attached to these manifold pipes, which can add up to an additional inch of space required between collectors. In most cases, you will need 3 to 4 inches between collectors when mounting. You also need to have about 12 inches of roof space on each end of the collector array for pipe runs and roof penetrations.

Draw a diagram of your collector array on paper and calculate all measurements. Now go up on the roof with your marking crayon, chalk line, tape measure and square. Lay out where the collectors will be mounted. Make sure you are not above the overhang with any part of the array. All collectors should be pitched slightly toward the supply end to facilitate system draining. This is critically important in drainback or draindown systems, where the pitch should be exaggerated. For all systems, the collectors should be pitched so that there is 1/8" of drop per foot of run. When installing a drainback or draindown system that uses just water as the solar fluid, you should increase this to 1/4" of drop per foot of run.

Mark the corners of the array and then snap a chalk line around its perimeter. When snapping the lines that are perpendicular to the roof rafters, pull the line tight and notice if the roof has developed sags. It is common for roofs sheathed with OSB or plywood to sag between rafters. If there is significant sagging, you will need to place shims between the roof and mounting brackets to ensure a proper installation, because collectors that are plumbed together must be kept in a straight line. If parallel collectors are not straight, getting them to seal together is difficult at best, and often impossible. Shims can be made of cedar or treated wood.

If there are multiple collectors in the array, start on either end and get the first collector mounted.

When starting the collector mounting process, first identify where the bottom mounting hardware will be located, and attach the mounting hardware in the appropriate manner. If the collectors will be flush-mounted, proceed to install the upper mounting hardware at this time as well. If the collectors are to be tilted at an angle greater than that of the roof, locate and install these mounting hardware feet as well. If you are inexperienced, mount the collector hardware one set at a time; after the first set is installed, proceed to mount the first collector. (Experienced installers often install all the mounting hardware sets before they attach any collectors.) When measuring for subsequent collectors in the array, don't forget to allow for the distance between the collectors. Leave the mounting hardware fasteners loosely attached to the collectors until all collectors are mounted, and then go back and tighten all the fasteners.

There are two methods of attaching collectors together. One uses brass unions; the other uses copper couplers. We prefer to use solid brass unions between collectors. The unions are soldered onto the collectors while they are still on the ground or in the shop. Some collectors come fitted with unions from the factory. The use of brass unions on the collectors minimizes the amount of soldering on the roof, and multiple collector arrays are much easier to install when unions are used. Also, if you ever have to replace an absorber plate, it is much easier if brass unions were used during installation. If you are using unions, just snug them up as you are mounting the collectors, and do the final tightening after all the collectors are in place. The best ones use an O-ring or gasket in the union.

Copper couplers are harder to use for several reasons. First, the copper collector nipples must be cleaned and fluxed before the collectors are slid together. Next, sliding the second and subsequent collectors together (necessary when couplers are used) is often difficult after the collector is already attached to the mounting hardware. If you attempt to slide the collectors together before fixing the collector to the mounting hardware, it is often very difficult to get the collector to line up with the mounting hardware. If you are using couplers, wait to solder them to the collectors until all the collectors are mounted and pushed tight together.

It is extremely rare to have the copper nipples line up perfectly when attaching collectors together in an array. Of course they will be close, but they need to be perfectly lined up when making the connections. Brass unions are slightly more forgiving. Note that the copper nipples are relatively soft and bend quite easily. If you notice a deformed nipple when uncrating the collectors, attempt to get it straightened before taking the collector to the roof. If couplers are used, the nipple must be perfectly round. You can purchase a tool from a plumbing supply house that will help you straighten out a deformed nipple. If you have unions attached to the nipples, you can tap the union halves with a rubber mallet to get them to line up. If you are using couplers, you can't tap them, so instead insert a broom handle or other suitable tight-fitting implement into the nipple and carefully bend it into position.

Be careful when bending the nipple that you don't deform it by making it out of round.

When plumbing flat plate collectors, it is important to have the supply piping attached to one of the ends of the bottom manifold and the hot return piping attached to the top manifold end on the opposite side of the collector. If you connect to two ports on one side, you will create uneven flow rate through the collector. If multiple collectors are used in an array, this same plumbing arrangement must be adhered to. You will want the array set up in a parallel design connecting the manifolds of each individual collector to make one large manifold.

In most cases you will want to connect all the bottoms together and all the tops together. This design will create an even flow rate throughout the array and will result in even heating. Even flow will also give the array maximum efficiency. It is also important to never exceed eight collectors in one array when using an internal manifold. This reduces damage that could be caused by expansion and contraction.

If you have to, you can make an external manifold. However, this system will have slightly lower efficiency because of heat loss through the exterior pipes. If the collectors are in a sawtooth configuration, an external manifold is usually necessary. For residential applications, you would never want to plumb your array in series. You can end up with water that's too hot and have decreased efficiency for the collectors on the end of the array. You would want to plumb the array in series only if you needed very hot water for commercial applications.

If the ridge of the roof runs east and west, a large flat surface will be facing south, which allows for easy collector mounting. If the ridge runs north and south or the roof is flat, a sawtooth array can be installed.

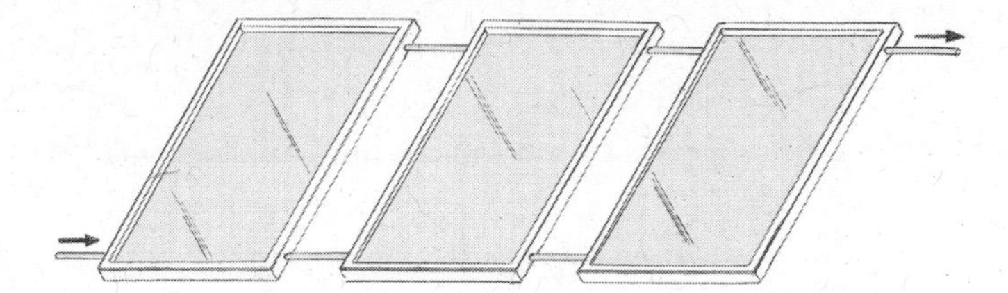

Figure 8.10: *Collector array in parallel with internal manifold*

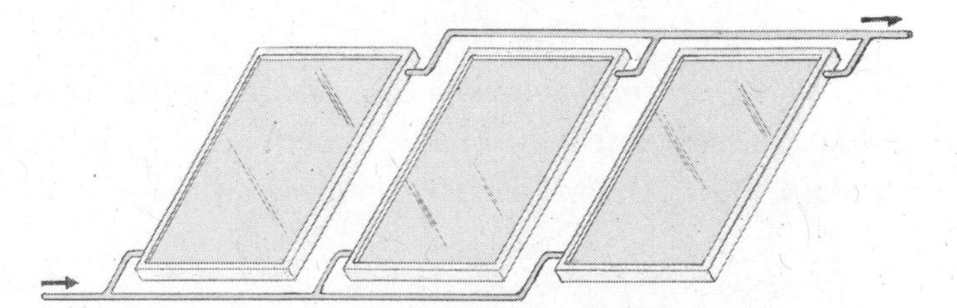

Figure 8.11: *Collector array in parallel with external manifold*

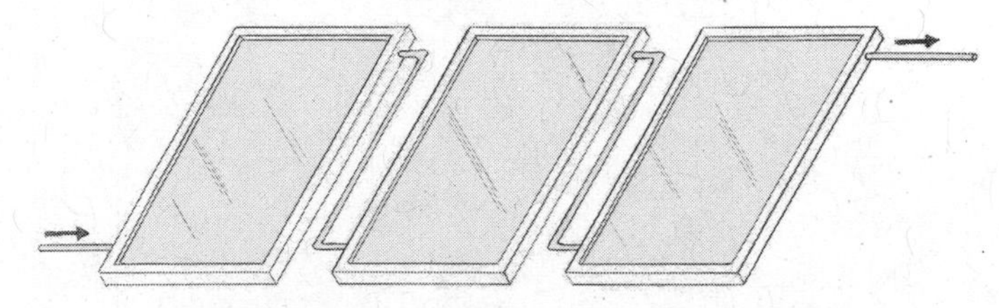

Figure 8.12: *Collector array in series*

(See Figure 8.13.) This configuration is also very common in large arrays that have multiple rows. For sawtooth configurations or arrays with multiple rows, the distance between collectors is critical. If the collectors are placed too close together, they can shade each other. You need to make sure that the collector in the front will not shade the one behind during the winter solstice.

Figure 8.14 can be used to calculate the distance necessary between collectors. For your latitude, find the appropriate roof pitch and collector-mounting angle. The number in the table will need to be multiplied by the vertical length of your collector. The result will be the distance from the front of the first row of collectors to the front of the second row. For example, if you live at 45°N latitude and want to mount your collectors at 45 degrees on a 4:12 roof, you would see that the multiplier is 1.563. If you are mounting a 4' x 8' collector vertically the front of the second collector would need to be 150" (96 × 1.563) behind the front of the second collector as measured up the slope of the roof.

Figure 8.13: *Sawtooth array*

Some people build a catwalk several feet below the collector array to facilitate easy access to the array. This may be desirable in high snow areas, but is usually not necessary. Typically, snow slides off collectors quickly, especially for heating systems where the collectors are mounted at a steep angle to maximize the winter sun. After a wet sticky snowfall, the snow can freeze to the collectors, especially if the temperature drops significantly after the snowfall. A catwalk can make scraping the collectors a lot easier. This is not absolutely necessary because even ice will melt off the collectors when the sun comes out, but one can get a head start on heating if you manually scrape off most of the snow early in the day after a snowfall.

If a catwalk is installed, make sure it is located well below the bottom of the collectors to avoid snow buildup. Remember that snow will pile up at the base of the collectors as it slides off them.

The feed-and-return pipe runs to and from the collectors typically have to penetrate the roof and extend into the attic. Once in the attic, the pipes can be run to a convenient place where they can then extend down to the basement or mechanical room where the storage tank is to be located.

Latitude		25			35			45			55		
Collector Tilt		15	25	40	20	35	50	30	45	60	40	55	70
Roof Pitch													
Flat	0 degrees	1.354	1.542	1.729	1.813	2.042	2.292	2.771	3.125	3.458	6.104	6.813	7.500
1:12	5	1.208	1.375	1.542	1.542	1.729	1.938	2.146	2.417	2.667	3.667	4.104	4.521
2:12	9	1.104	1.271	1.417	1.375	1.542	1.729	1.833	2.063	2.271	2.792	3.125	3.438
3:12	14	1.021	1.167	1.292	1.208	1.375	1.542	1.542	1.729	1.917	2.167	2.417	2.667
4:12	18		1.083	1.229	1.125	1.271	1.417	1.417	1.563	1.729	1.854	2.063	2.271
5:12	23		1.021	1.146	1.021	1.167	1.292	1.229	1.396	1.542	1.563	1.750	1.917
6:12	27			1.104		1.104	1.229	1.146	1.271	1.417	1.396	1.563	1.729
7:12	30			1.063		1.063	1.188	1.083	1.208	1.354	1.313	1.458	1.604
8:12	34			1.042		1.021	1.125	1.021	1.146	1.271	1.188	1.333	1.479
9:12	37			1.021			1.104		1.104	1.208	1.125	1.271	1.396
10:12	40			1.000			1.063		1.063	1.167	1.083	1.208	1.333
11:12	43						1.042		1.021	1.125	1.021	1.146	1.271
12:12	45						1.021		1.000	1.104		1.125	1.229
13:12	47						1.021			1.083		1.083	1.208
14:12	49						1.000			1.063		1.063	1.167
15:12	51									1.042		1.042	1.146
16:12	53									1.042		1.021	1.125
17:12	55									1.021		1.021	1.104
18:12	56									1.021			1.083
19:12	58									1.000			1.083
20:12	59												1.063

Figure 8.14: *Inter-row spacing multiplier*

The best way to penetrate the roof is to use a roof boot. These items are commonly used to make a waterproof roof penetration for sewer vent pipes, which are similarly sized to insulated solar piping. (An insulated ¾-inch or 1-inch copper pipe will be 3 to 4 inches in diameter when insulated.) The roof boot is usually made of an aluminum plate, with a neoprene collar mounted in the center of the plate. The insulated pipe fits tightly through the neoprene collar and requires no additional sealing. Some roof boots are made entirely of plastic. You will need one roof boot for each pipe. Typical roof boots are self-adjusting for any standard roof pitch. They are available at most lumber yards. Be sure to install them properly by fitting the top of the roof boot below the upper shingles and over the shingles below it. Use liberal amounts of roofing cement when installing roof boots.

Some hardware mounting methods are also designed to be used with roof boots. Several companies offer a product with a round post that extends off of the roof. There is a threaded hole in the center so you can make a watertight connection to a rail or beam that runs the length of the collector array. This post and rail configuration not only allows for better leak protection but also allows the fasteners to be directly lagged into the roof rafters or trusses. In situations where the underside of the roof is not accessible, such as a cathedral ceiling, this may be the only way to mount your array.

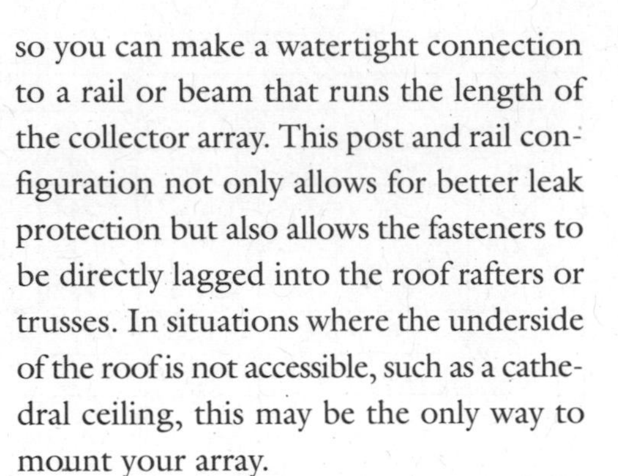

Figure 8.15: *Roof boot*

Figure 8.16: *Post and rail mounting*

Ground Rack Mounting

When there is not an appropriate roof to mount the collectors on, a ground rack mounting system can be used. Solar thermal collectors, whether flat plate, ICS, evacuated tube or concentrating, all have ridged frames. These frames can be bolted to a framework, resulting in a safe, durable and long-lasting installation.

It is extremely rare to use ground mounts for any type of solar water heating or space heating system except pressurized

antifreeze systems. Ground-mounted arrays are very popular and offer a major advantage over other mounting systems in having a greatly expanded number of collector location options. The location of the ground-mount rack should be identified using the criterion specified in the previous section on siting the collectors. You want to locate the rack as close to the building as possible to minimize the length of the pipe runs. Note that the piping between the collector array and the house will probably be buried, so note any obstructions that would interfere with trenching, such as a driveway or buried power lines. Always know what may be buried in the yard before you dig.

If you live in an area where snow accumulates, the rack should be designed so the bottom of the collectors is raised above the ground. The bottom of the collectors should be at least twice as high above the ground as the deepest snow you would experience during an average winter. This height is needed because snow will slide off the collectors and make a pile at the base of the array. You may still need to compact the snow after a heavy snowfall, but this would be a rare occasion.

Ground-mount collector racks can be built out of wood or metal. Because the collector array is subject to potentially high winds, the rack must be constructed to withstand the highest winds that normally could occur in your area. In areas subject to possible high winds, such as coastal or mountainous areas, concrete footings should be provided to anchor the rack to the ground.

Footings should extend below the frostline. Concrete footings should be placed on undisturbed soil. The footing piers should have a framing member embedded in the cement or have an anchor bolt embedded at least four inches into the top of the pier. Where high winds are not

Bob Ramlow

Figure 8.17: *Ground-mounted collectors on concrete pillars*

experienced, wooden piers can be sunk into the ground below the average frost depth. Treated four-by-four wood works well.

All collector manufacturers have collector mounting systems. These mounting systems include tilt mounting hardware or tilt racks. This hardware allows for easy collector mounting on any pitched roof or flat surface. The best way to ground mount solar collectors is to create two parallel rails and fasten the flat roof-mounting system to those rails. The distance between the rails is determined by the specifications provided by the collector manufacturer. This would be the distance between the front feet and the rear legs. Parallel rows can be made of treated wood or metal rails.

The ground-mount racks should be strongly built with adequate crossbracing. Racks can be constructed of angle iron, galvanized pipe or wood. Remember to isolate different types of metal with rubber, neoprene or other nonconductive washers to minimize galvanic reactions between dissimilar metals. Aluminum angle iron is an excellent material to construct the racks with. If galvanized metal is used, be sure to properly paint all surfaces, especially the cut ends. Likewise, all wooden racks should be painted to extend their

MIDWEST RENEWABLE ENERGY ASSOCIATION

Figure 8.18: *Installing collectors on a ground mount*

lifespan. Remember that these systems will last the life of the home they are servicing, so build accordingly.

We have seen several homeowners incorporate a ground mount into a garden shed or storage shed. One customer used the area under the array to store firewood. Another used the building as a kids' playhouse. In these cases, an actual roof was built on the rack and the collectors were mounted flush on the roof with standard collector mounting hardware. A roof was also built on the back side of the array to make an A-frame structure. Add two end walls and presto, you have a shed for little more cost than that of the rack. Be creative!

Air Collector Mounting

Because air-type solar heating systems are intended to operate only during the winter months, when the sun is low in the sky, the collectors should always be mounted at least at a 60-degree angle or greater. Most air-type collectors are mounted at a 90-degree angle. Wall mounting has been most popular. The vertical mounting of the collectors takes advantage of snow to reflect additional solar energy into the collector for increased efficiency.

Mounting the Heat Exchanger

At any time during the installation process you can place the solar storage tank and start plumbing the heat exchange module. If you are using a prepackaged heat exchange module, this part of your installation could be as easy as setting the module on your storage tank and tightening a few unions. If you are making your own heat exchange module with individual components, make sure your gauges and valves are visible and easily accessible.

Heat exchangers are mounted to storage tanks in several ways, depending on the type of tank. Most storage tanks made specifically as solar storage tanks have additional ports on the sides or top. It is also common to modify an electric water heater for use as a solar storage tank.

The first step is to set the storage tank in its appropriate location. If a thermosiphon heat exchanger will be used, it is best to elevate the tank by six inches. Since the coldest fluid will be dropping to the lowest place, this will assist in the natural convective heat transfer. Using a storage tank specifically designed for use as a solar storage tank works very well. Such tanks typically have four ports on the top, plus the T&P valve located either on the top or on the side near the top. The ports are usually clearly marked as "Hot," "Cold," "Solar In" and "Solar Out" and are typically ¾-inch national pipe thread (NPT). The Hot port is attached to the backup water heater. The Cold port is where the cold supply line is attached.

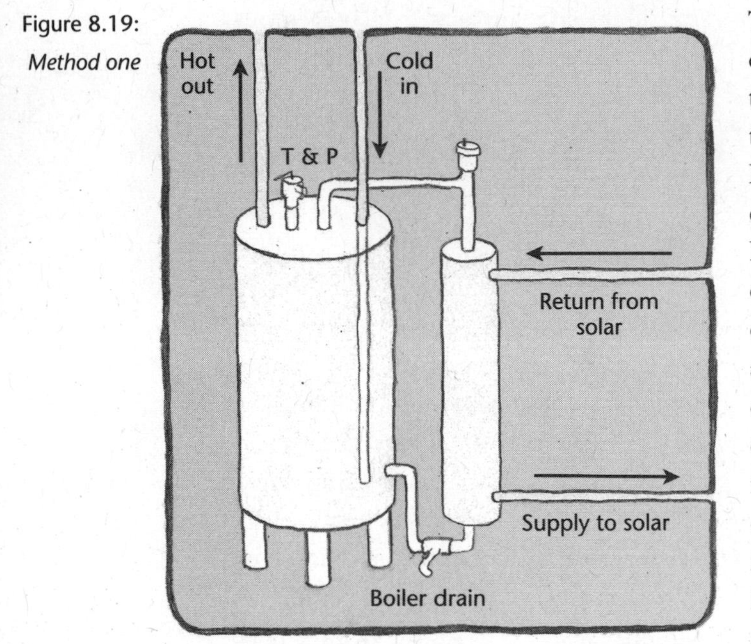

Figure 8.19:
Method one

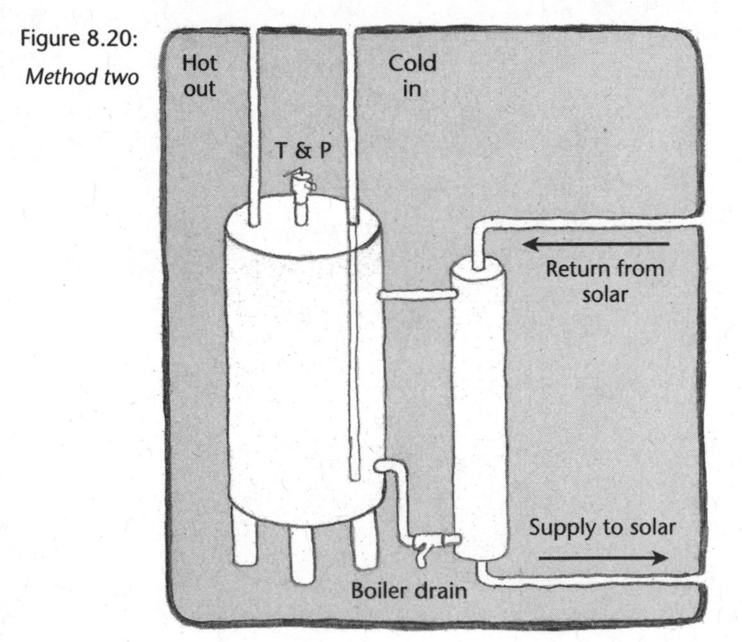

Figure 8.20:
Method two

The Solar In port is attached to the top, or hot side, of the heat exchanger, and the Solar Out port is attached to the bottom or cold side of the heat exchanger. Figure 8.19 shows a thermosiphon heat exchanger mounted on a solar storage tank. Note that when using a thermosiphon heat exchanger, the Solar Out port on the top of the tank is plugged and the drain is attached to the bottom of the heat exchanger. Whenever a thermosiphon heat exchanger is used and the hot water pipe from the top of the heat exchanger must enter the top of the storage tank, it is imperative that an automatic air vent be installed at the high point of this pipe run as shown in Figure 8.19.

Using a new electric water heater (or one in very good condition) as a solar storage tank is a common practice. They are easily available, are of high quality and are a good value. These tanks usually have two or three ports on the top—a hot port that is open to the top of the tank, a cold port that has a dip tube fitted to it, and a temperature and pressure-relief valve port. If this valve port is not on the top, it will be on the side near the top. A drain valve with a boiler drain is at the bottom of the tank, and one or two electric elements are screwed into the side of the tank. All the ports have a ¾-inch national pipe thread except the electric element ports, which are usually 1-inch machine thread.

If you are using a new electric water heater as your solar storage tank, you can use the top electric element port on the side of the tank as the "Solar in" port from the top of the heat exchanger as shown in Figure 8.20. If you remove an element, you can see that the threads on the element are not tapered, and at the end of the threads there is a flat faceplate where an O-ring or gasket is fitted. When the element is screwed tightly into a water heater, the faceplate comes into contact with the water heater flange and the O-ring seals the fitting to the tank.

To make this port useable with tapered pipe thread, a modified 1" × ¾" brass bushing is used. You will need to have a machine shop modify a standard 1" × ¾" brass bushing by turning the end of the nut flange flat on the inside where the threads taper into the nut. This creates a flat surface where an O-ring can be fitted, as the element is set up for. With this modified bushing screwed into the tank, you have a new port at a perfect location to attach the top pipe from a thermosiphon or pumped heat exchanger. It is best to use the drain port to attach the bottom of the heat exchanger to the tank. Remove the original tank drain and install a ¾" × 3" brass nipple in the port. Then install a ¾" × ¾" × ¾" brass tee on the nipple. On one branch of the tee, install a ¾" brass boiler drain. The other port goes to the heat exchanger.

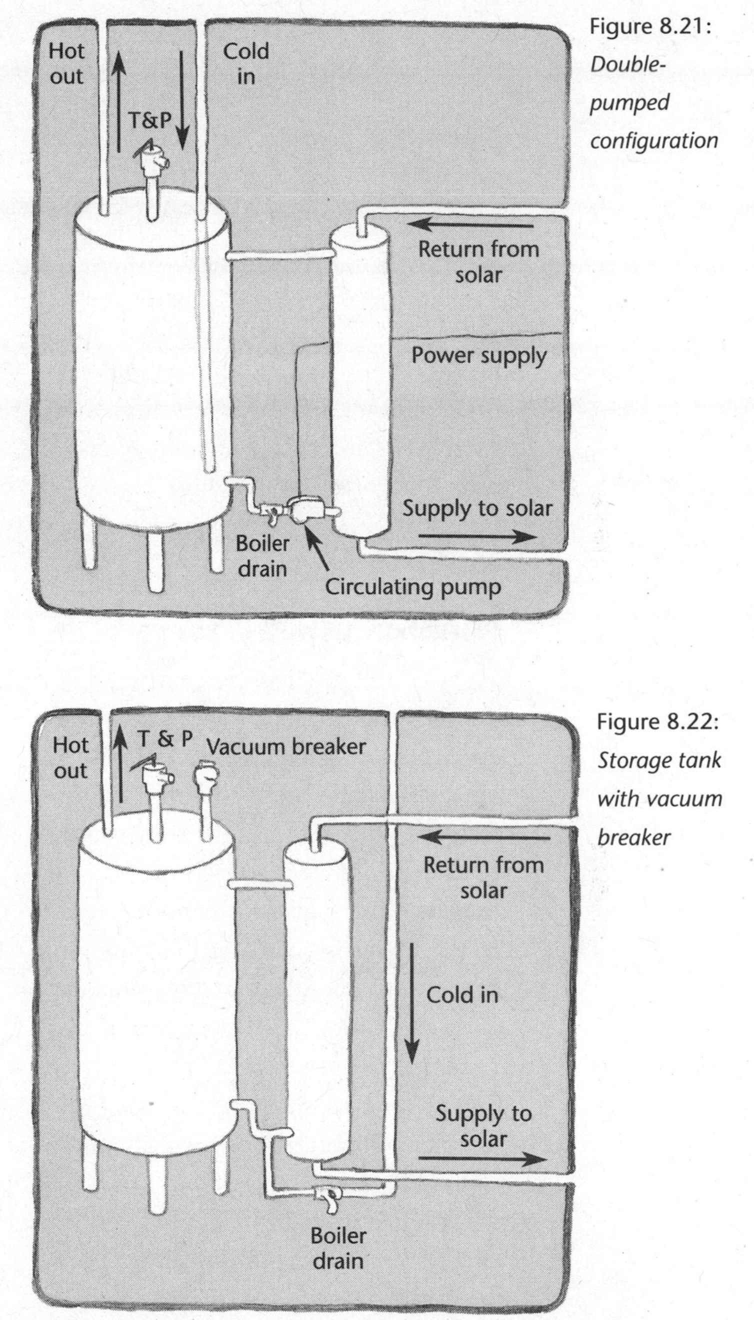

Figure 8.21: *Double-pumped configuration*

Figure 8.22: *Storage tank with vacuum breaker*

If you are placing a second pump on the system, you will need to install that prior to mounting the heat exchanger. Figure 8.21 provides the location of the secondary pump. Ensure that this is a small pump to maintain stratification of the tank.

If you are using a plastic-cased fiberglass tank, you will need to install a vacuum breaker on the system, eliminating one of the ports on the top of the tank. You will need to install the cold inlet on a brass tee near the bottom of the heat exchanger. Figure 8.22 shows the new location of the cold inlet.

Plumbing the System

Many of the problems encountered during the installation of a solar water heater occur because of poor workmanship while plumbing. The most common of these results from neglect while soldering the copper tubing. A soldered copper joint can last more than 100 years if done correctly. If this is your first time soldering copper, we suggest that you practice a couple of times before working on your system. We should also stress that the instructions on soldering copper tubing must be adhered to for the joint to seal and last.

When retrofitting a home with solar energy, often the most difficult part is trying to find a way to get the pipes from the collector to the solar storage tank without having to tear out some Sheetrock. There are a number of options to consider before that becomes necessary. Always start by looking for some sort of chase or recess in the wall that leads directly from the utility room to the attic. Often a home will have an enclosed chimney with a little extra space, or chase, around the main sewer vent. This is always the best place to start, but we should warn you that unless you can see all the way through, there might be obstructions. Sometimes it may appear that you can make it from top to bottom, but when you try to shove some copper pipe through, you realize that the space isn't big enough or that the floorboards butt tight against the chimney. If there isn't a chase, the next best thing to look for is closets that are stacked upon one another. Most people don't mind a couple of pipes hidden behind their clothes. If the attic is only one story above the utility room, this is usually your best bet; it will sometimes work on a two-story home if you get lucky. If you are completely adverse to the idea of opening your walls, the last option is to make an exterior pipe run. You should consider this only if you are willing to insulate your pipes well, especially if you live in a cold climate. Exterior pipes can often be hidden behind a gutter drain to retain the aesthetics of the home. We have also seen installations where the pipes were enclosed in a box made of the same materials as the siding. This is probably the best way to make an exterior pipe run because you can add some extra insulation inside the enclosure.

We have already recommended that you make a simple line drawing of the pipe runs. This will give you an idea of the amount of copper tubing and the number and type of fittings you will need for the job. If you are going to be making a long pipe run, you will have to be able to slide the pipe in from either end and will need the space to do so.

Installing and insulating the pipes is the most time-consuming part of most installations. The solar-loop piping and pipe insulation can also be expensive. The best location for the pipe runs is a tradeoff of pipe run length and ease of installation.

Once you have determined the location of the pipe runs, you need to gather all the necessary materials:

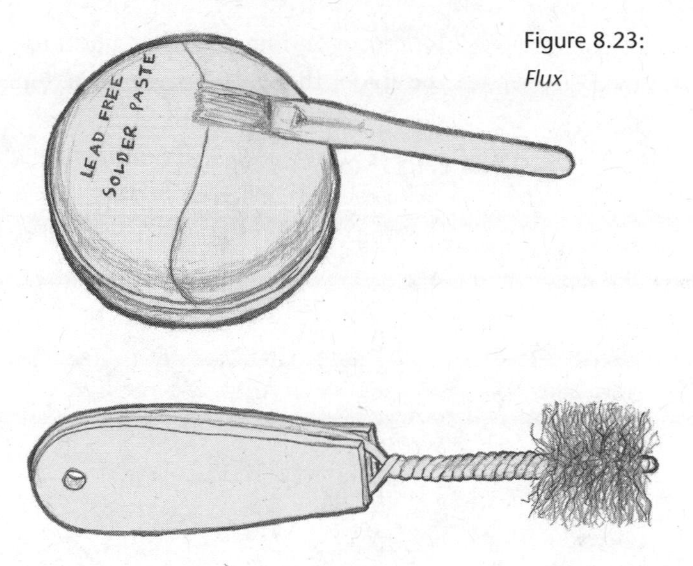

Figure 8.23: *Flux*

Figure 8.24: *Wire brush*

1. Torch. Torches come in a variety of models, but most of the newer ones have a built-in igniter. With this type you only need to turn on the gas and click the trigger. With the older models, you will have to ignite the flame manually.
2. Gas. The most common type of gas used is propane. Propane works fine for soldering, but you can use acetylene or Mapp gas if you already have that available. Acetylene and Mapp tend to burn hotter than propane and will cut down the amount of time you spend heating the copper. However, they increase the risk of overheating. If you are new to this, the safest bet is to start with propane. Once you have gained a little experience you can move on to Mapp gas.
3. Solder. Ensure that you are using a lead-free solder. Even though potable water may not be traveling through the solar loop of the system, it is never a good idea to bring lead into your home. We recommend using the 95/5 lead-free solder that comes on a roll. It is composed of 95 percent tin and

5 percent antimony. You can also use the solder that is 100 percent tin, but the 95/5 is slightly stronger. Either type is suitable. You shouldn't use a paste solder or a flux and solder mixed together, as your joints will not seal as well.

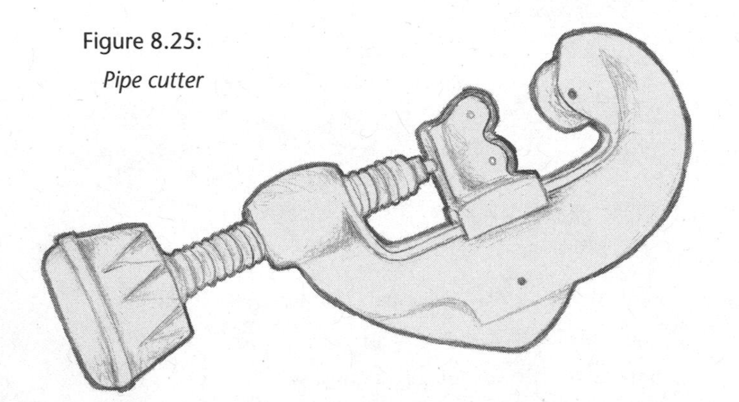

Figure 8.25: *Pipe cutter*

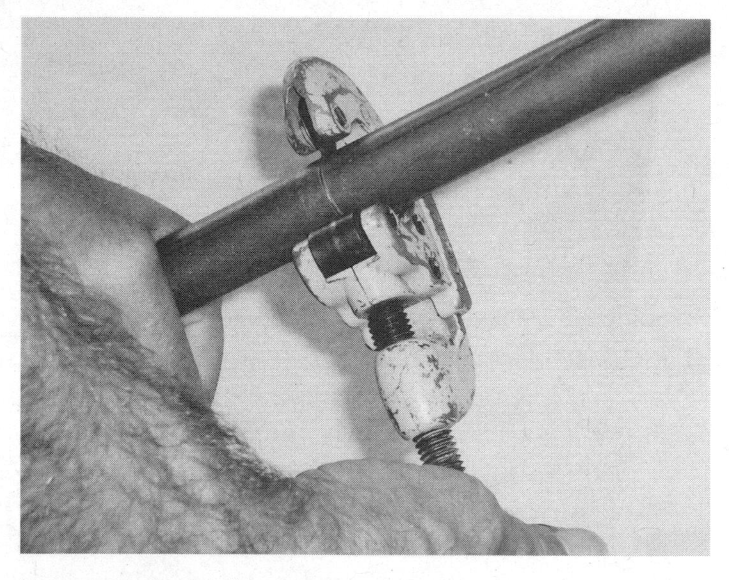

Figure 8.26: *Cutting the tubing*

4. Flux. Flux acts as a cleaning agent for the tube and the fitting. We highly recommend a water-soluble flux because it will be easier to clean from the inside of the pipes when the system is flushed.
5. Wire brush.
6. Grit cloth. Don't use emery cloth. Many types of emery cloth contain oil that may degrade the joint. If you plan to use emery cloth, ensure that it is oil-free.
7. Wet rag.
8. Gloves.
9. Measuring tape.
10. Pencil or marker.
11. Insulated heat mat or small sheet of metal.
12. Copper pipe and fittings.
13. Pipe hangers and fasteners.
14. Pipe cutter or hacksaw.

Once you have gathered everything you need, you can begin to assemble the pipe runs. It is best to start from one end, either top or bottom, and work the other way. If two of you are working together, make sure you will meet in the same place. Since you already have the pipe runs planned out, you know where the fittings will need to go. Follow the next steps to ensure well-sealed joints.

Step One: Measuring and Cutting the Copper Tubing

From your starting point, measure the distance to the next fitting. You can either hold the copper pipe in position and mark the distance, or you can determine it with a tape measure. Most likely, you will use a combination of both. With either method, be sure to take into account the length inside the fitting.

You have a couple of options for cutting the tubing. We recommend using a pipe cutter like the one seen in Figure 8.25. You will need to place the circular blade at the appropriate length and hand tighten it to the tubing. Spin the pipe cutter around a couple of times until you start to feel it loosen. Retighten the blade and repeat until the tube is cut. When using a pipe cutter, lead with the wheels instead of the blade. You can also use a hacksaw or a reciprocating saw with a metal-cutting blade, but we discourage these methods because they do not always make a square edge and often leave metal burrs and chips on the end of the tubing. If you cut this way, you will want to use a metal file to clean off all the burrs.

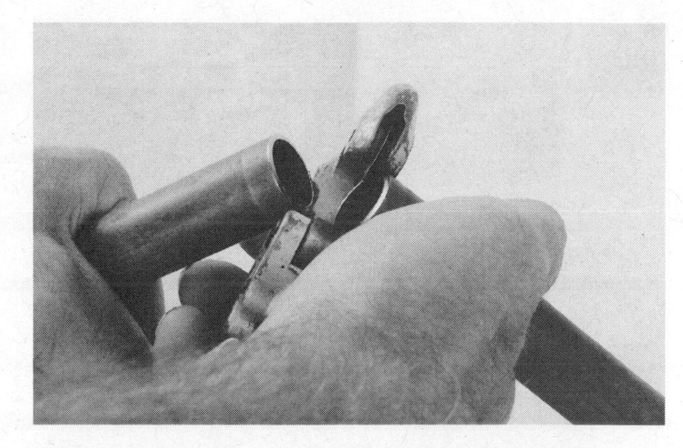

Figure 8.27: *Separating the tubing*

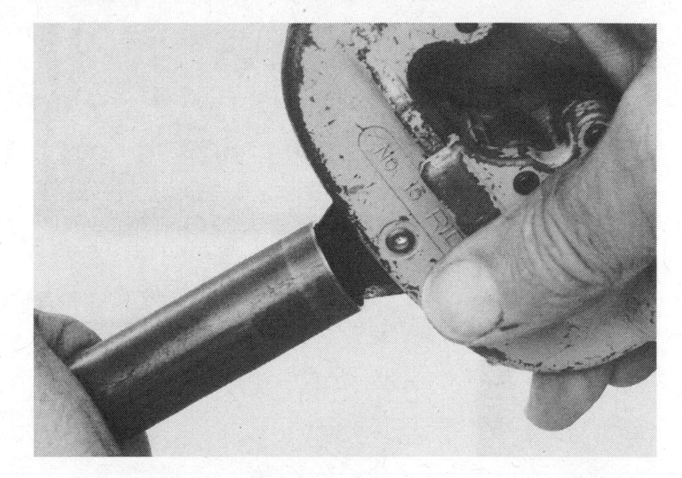

Figure 8.28: *Removing the burrs from the inside of the pipe*

Typically, the quickest way to install the tubing is to cut and dry fit all the pieces together first, and then go back and solder them all together. If you are working in a tight space, such as an attic, you may want to work through all the steps instead of dry fitting all the pieces.

Step Two: Clean the Tubing and Fittings

Thoroughly scrub the outside of the pipe and the inside of the fitting with either a wire brush or grit cloth. We have found it best to use the wire brush on the inside

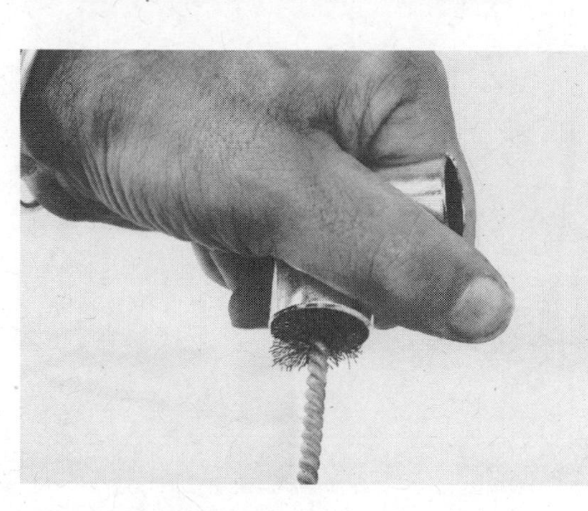

Figure 8.29: *Cleaning the fitting with a wire brush*

of the fitting and the grit cloth on the outside of the pipe. Either one will work fine, but using both makes it a bit easier. Cleaning the copper removes the oxidation and makes for a pure surface for the solder to adhere to. If you don't properly clean the copper, you will undoubtedly have leaks. The most common reason for leaky joints is skimping on this step. If you are soldering a lot of joints, you can purchase a wire brush that will attach to an electric drill to save some time. You will also want to scrape away any burrs that have attached to the inside of the pipe, using the attachment on the pipe cutter, a knife or a deburring tool. A deburring tool will do the best job.

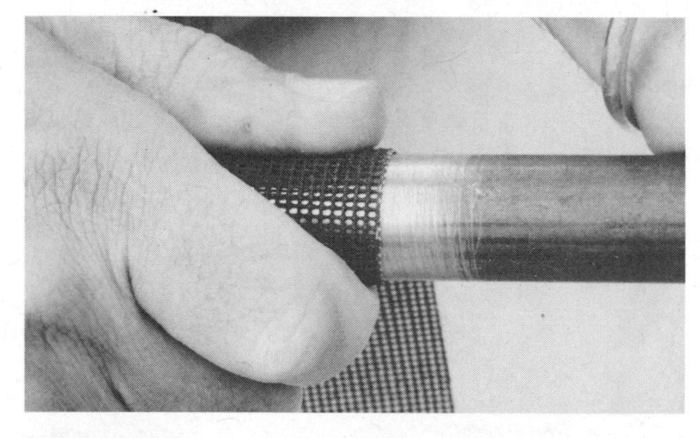

Figure 8.30: *Cleaning the tubing with grit cloth*

Step Three: Applying Flux

You cannot solder without flux. The solder won't adhere and will simply slide off the pipe. Most flux comes with a small fine-bristled brush. This is the best way to apply the flux to the copper. You will want to apply flux to both the outside of the tubing and inside of the fitting. Brush flux on all of the copper that will be inside the joint. You only need to cover the copper completely. You don't want gobs of excess flux remaining inside the pipe. Flux is acidic and may degrade your solar fluid if too much is left in the tube. Once the flux has been applied, slide the tube into the fitting.

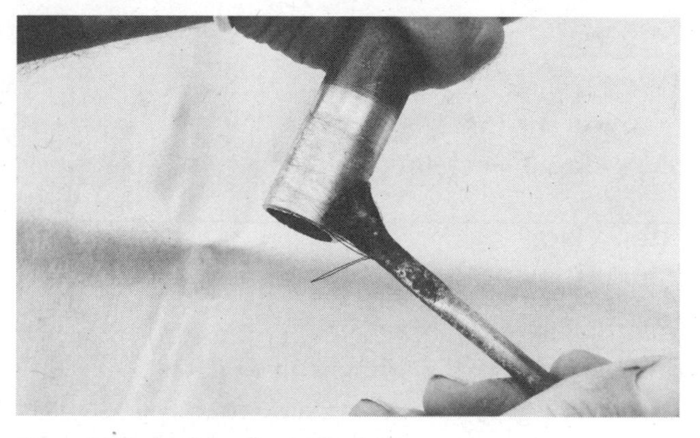

Figure 8.31: *Applying flux to the tubing*

Step Four: Apply Heat to the Joint

Ignite your torch and use it to apply heat to the joint and tube. You will want to heat the entire fitting and part of the tubing, hitting the flux directly with the flame. Try not to heat the place where the fitting overlaps the tube. When the flux starts to sizzle and boil, you know that your joint is hot enough. This process usually takes about 15 to 30 seconds when using propane. Don't overheat the joint. If the joint gets too hot, it may burn the flux and the solder will not adhere, or the burnt flux will block the flow of the solder.

You need to use caution when using a torch in tight areas. Often pipe runs will go through attics and crawl spaces without much room to maneuver. If you have to apply heat to a joint that is next to a stud or something else flammable, be sure to place an insulated heat pad or a sheet of metal behind the fitting. Some insulated heat pads will have eyelets so you can hang them on a nail to keep your hands free.

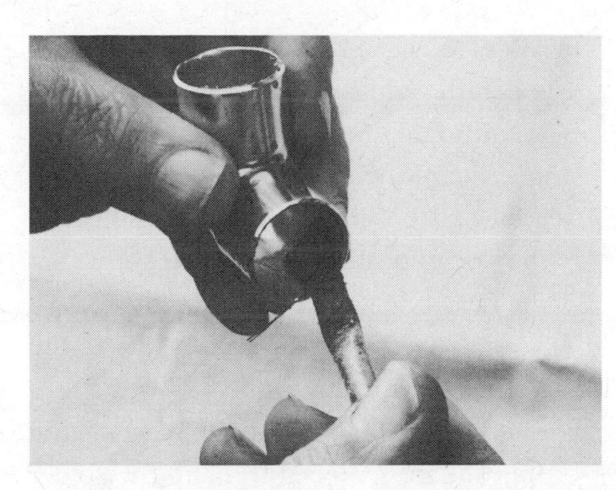

Figure 8.32: *Applying flux to the fitting*

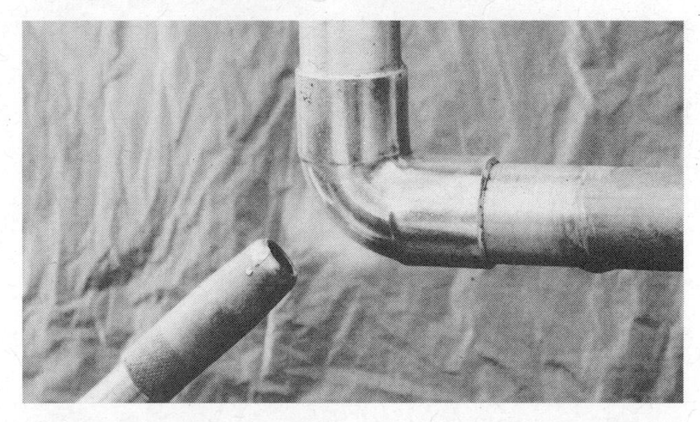

Figure 8.33: *Heating the fitting*

Figure 8.34: *Applying solder to the joint opposite the location of the torch*

Step Five: Applying the Solder

Unravel a couple of inches of solder from the roll and have it ready before you start applying the heat to the joint. You will use about the same amount of solder as the size of your pipe. For instance, for a half-inch pipe, you will use half an inch of solder. Next, take your roll of solder and touch the end to where the tubing meets the fitting. Direct the solder to the opposite side of the pipe that you are applying the heat to. When you apply the solder to the side opposite the heat, you know that the near side will be hot enough.

If you start applying the solder to the side near the flame, you will not be sure that the opposite side has reached the appropriate temperature, and the solder may not fill that space. If your joint is hot enough, you will see the solder melt and be sucked into the joint. Even if the joint is placed vertically, you will see the heat's capillary action draw the solder upwards. Once the solder starts flowing into the joint, you should take the heat away from the joint so as to not overheat anything.

Step Six: Don't Touch It

If the joint is rattled too much while it is still hot, the solder may not cool evenly, which may lead to leaks in the pipe run. Some plumbers will use a wet rag to wipe the joint while it is still cooling, but this is not necessary. You can still wipe excess flux from the joint once it has cooled. Wiping the excess will prevent the greening of the pipe over time.

Step Seven: Secure the Tubing

It is important to securely attach all your piping to a wall or ceiling. A popular method professional plumbers use is to anchor a Unistrut channel to the wall and use proprietary clamps that fit into the channel and clamp to the pipe. These pipe-hanging systems make for a clean and solid installation. You can also use bell hangers or C-shaped clamps that nail or screw to the wall. It is important not to attach a circulating pump, especially a PV-powered pump, directly to an interior or a hollow wall because the vibrations generated by the pump will amplify in the wall and can become a nuisance. When stabilizing the piping near a pump, be sure to use rubber isolation clamps to minimize the noise.

When installing the pipe runs from the heat exchange area to the collectors, the piping needs to be supported every five feet of horizontal run. Many brands and types of pipe hangers will work just fine for this task. This is most important in drainback and draindown systems, where any sags in the pipes could allow water to settle in the sags, freeze and burst the pipe where the pipe runs through an uninsulated area. Another potential problem is that water collecting in sags can inhibit the quick drainage needed with these types of systems.

With ground-mounted arrays, the insulated piping is usually run underground from the array to the building. The trench should be at least 18 inches deep, and 36 inches deep under driveways in northern climates where the ground freezes solid during the winter. An important consideration when burying copper pipes is to compensate for the expansion and contraction of the piping. Copper piping will expand and contract considerably when it goes through its daily heating and cooling cycle, especially in cold climates where the temperature of the pipes can vary by more than 200 degrees each day. When pipes are buried, tightly packed ground can prevent them from expanding when they warm. This can cause buckling of the pipes and eventually cause leaks to form. This phenomenon is especially evident in long, straight pipe runs.

There are a number of things you can do to minimize potential problems. The most important consideration is to use type K soft copper pipe for all buried piping. This type of copper pipe is relatively flexible and is somewhat self-compensating. Another advantage of soft copper pipe is that it comes in long rolls, usually 60- or 100-foot continuous lengths, which minimizes the number of joints that will be buried. It can also be bent, which minimizes joints by eliminating the need for elbow fittings.

The best way to protect this pipe is to place it inside a PVC drainpipe. We like to use a 4-inch PVC pipe for each pipe. The 4-inch size is easy to slide over the pipe after it has been insulated, and it provides plenty of room for expansion and contraction. Some contractors prefer to place both insulated pipes inside a single, larger PVC pipe. A 6-inch inside diameter (ID) pipe is a tight fit. Be sure to glue all joints to keep them waterproof. If an elbow is needed in the underground pipe run, you can make the job of installing the PVC elbow easier by cutting the elbow in half lengthwise with a hacksaw, placing the halves together over the bent pipe and then gluing the PVC elbow back together and to the straight lengths on either side. It may make it easier to stuff the insulated copper pipe into the PVC pipe and glue it all together before dropping the whole thing into the trench.

Figure 8.35: *Attaching the pipe to the bell hanger*

Step Eight: Insulate your Pipes

All pipes should be insulated. Exterior pipe runs should be especially well insulated. Use a type of pipe insulation that will stay flexible after being exposed to the high temperatures experienced with solar water heaters. Requirements for pipe insulation have been covered in Chapter 4. When installing pipe insulation, be sure to miter your corners to ensure insulation thickness over the joint. When insulating a tee in the pipe, cut a notch out of the long piece at the joint, and cut the end that butts against the notch into a point to fill the gap.

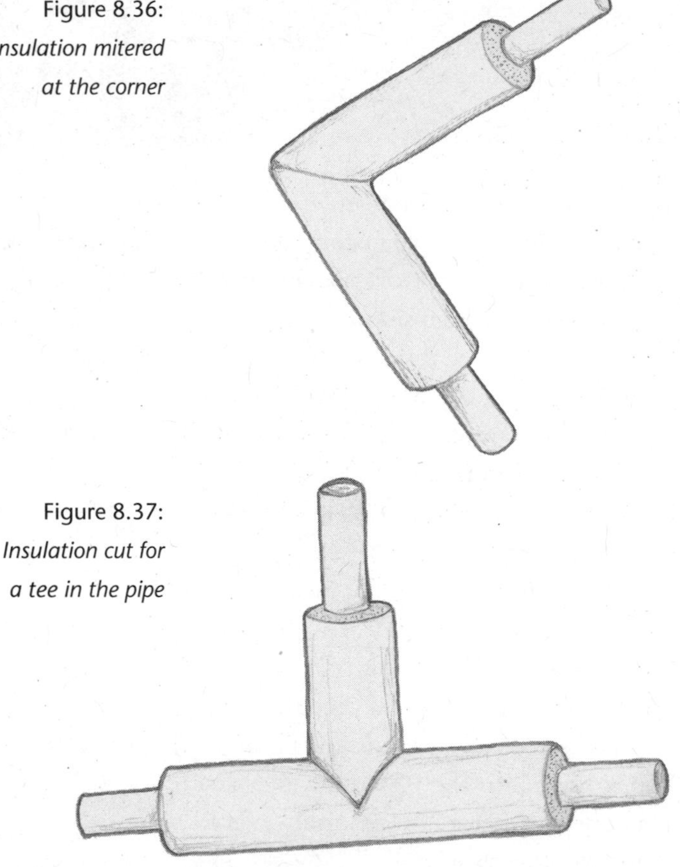

Figure 8.36: *Insulation mitered at the corner*

Figure 8.37: *Insulation cut for a tee in the pipe*

Testing the System

Once the whole system is plumbed, it is time to check for leaks. There are two ways to check for leaks: one uses air pressure; and the other uses water pressure. Using air pressure has two advantages: dry joints and no mess. To pressure test a system with air, you need to attach a fitting to the system to allow you to add air pressure to the piping loop. You can use a bushing that has a Schrader valve installed on it. You take one boiler drain out of the solar loop and screw the bushing/valve into that fitting. Some of these have hose thread, so you can just attach it to the end of the boiler drain.

With your portable compressor, start filling the system with air. You can monitor your progress by watching the pressure gauge. Pressurize the system to at least 50 pounds per square inch. Once it's pressurized, watch the gauge to see if the pressure is dropping. If there is a big leak, the pressure will drop noticeably right away. Note that if you are checking during the daytime, the pressure will fluctuate because the sun on the collectors will heat the air in the system, causing the pressure to rise, and when the sun leaves the collector or goes behind a cloud, the pressure will drop a little. If you have a leak and you have plenty of pressure in the system, the air escaping will create an audible noise, usually a whistling sound. Go over the whole system listening for a noise.

During the soldering process, you will have noticed that the solder filled the joints in a predictable and consistent manner. If you have a fitting that did not seem to take the solder properly, it is a good idea to not insulate that joint till after you pressure test the system. If you have a ground mount that has buried piping, you should do the pressure test before you backfill the trench.

You can pressure test the system with water too. The advantage of using water is that you should rinse the system anyway before you put the solar fluid in. By rinsing the pipes, you remove any leftover flux, which can degrade the solar fluid. The trouble with using water is that if you have a leak, the water may damage the building, especially if the installation is a retrofit and the building is finished. It is very important to have the collectors covered when using water to test the system, so please read the cautions outlined in the section on charging a system in the following chapter.

After the system is pressure tested, you can proceed to charge it with solar fluid as outlined in the section on charging a pressurized system. Make sure to remove the handles from the charging valves and hang them in a convenient but out-of-the-way location, and install caps on all drain valves.

Installing Solar Pool Heating Systems

Because solar pool heating systems are so common and mainstream, manufacturers offer very detailed installation instructions. Please read your instructions carefully before you begin. These instructions contain many tips that will not only save installation time but additional costs as well. Following these instructions will also help ensure proper performance. All the larger manufacturers also offer technical assistance for proper system design and component choices.

When installing these systems' collectors and piping, it is important to mount everything to facilitate proper drainage. The collectors must be angled so they drain, and all the piping should be mounted to drain. Always follow manufacturers' instructions when installing these systems.

When you order your collectors, you can always order the balance of the system's parts from the same vendor. They often come as a kit. Use black, schedule-40 PVC pipe and black, schedule-40 PVC fittings if possible for all your piping.

Make sure you get solid PVC pipe, not foam-core PVC. If you can't get black PVC, install white, schedule-40 PVC pipe and let it age in the sun for a year or clean the pipe with PVC joint cleaner before attempting to paint it. Black piping can significantly improve the performance of your solar heating system. Anchor the PVC piping every five feet on horizontal runs and every ten feet on vertical runs. Make sure the clamps you use on the PVC pipe are oversized to allow the pipe to expand and contract.

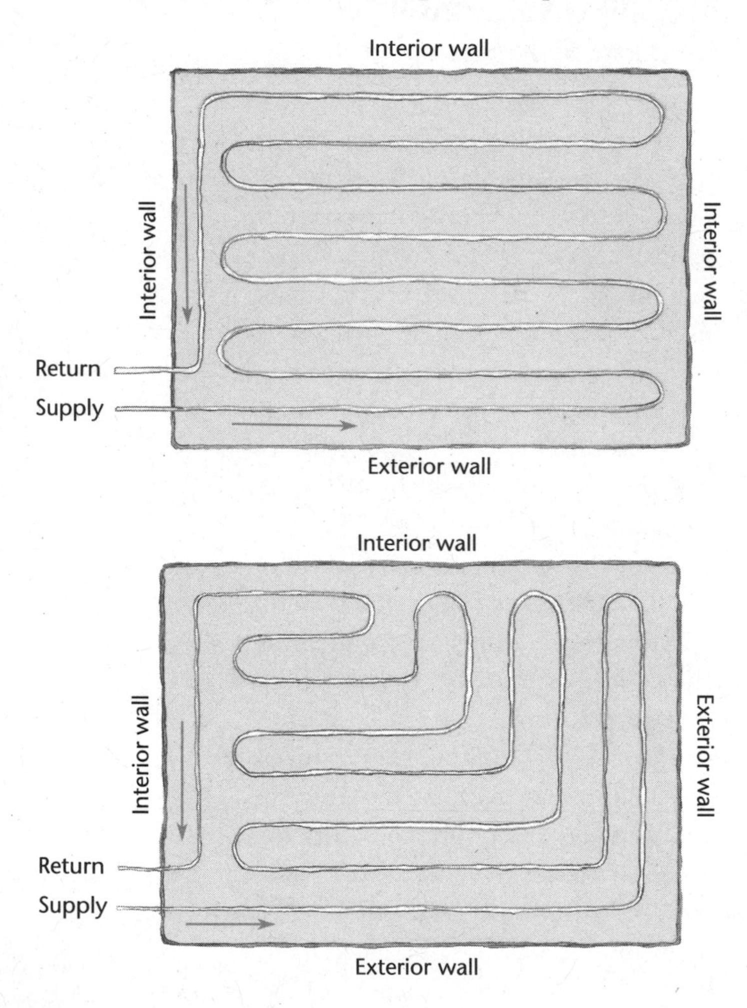

Figure 8.38: *Single-wall serpentine*

Figure 8.39: *Double-wall serpentine*

Installing Radiant Floor Tubing

Radiant floor tubing can be installed in either a dump system or a system with water storage. Both types will follow similar rules for installation.

Tubing Layout Patterns

The size of the tubing used in a solar radiant floor heating system will determine the length of each tubing circuit. The smaller the tubing, the shorter the length of the circuit. As heating fluid enters the tubing and flows through the circuit, the fluid cools down. Therefore, introducing the warmest heating fluid where the heat is most needed can maximize the heating system. This is typically along outside walls. Tubing spacing will also affect the performance of a radiant floor heating system. The closer the tube spacing, the more heat will be given off. Tubing is usually spaced closer together along outside walls, where more heat loss occurs, and farther apart in the center of a building.

Four basic patterns can be used, depending on how many outside walls, if any, are in the area to be heated by a circuit. They all follow the same basic principle: send the warmest fluid to the locations with the most heat loss first. This will create even heating throughout the area.

Single-Wall Serpentine

When one outside wall represents the major heat loss of a circuit, the single-outside-wall serpentine design is recommended. As shown in the diagram, the hottest heating fluid is directed to tubing laid along this outside wall where the heat is needed most.

The tubing then spirals to the area where heat is needed least. The tubing spacing should be closer where the heat loss is the greatest and can be wider for the majority of the circuit.

Double-Wall Serpentine

When two adjacent outside walls represent the major heat loss of a circuit, this pattern is recommended.

Triple-Wall Serpentine

When three adjacent outside walls represent the major heat loss of a circuit, this pattern is recommended.

Counterflow Pattern

Use this pattern when the heat loss of the circuit is evenly distributed throughout the circuit.

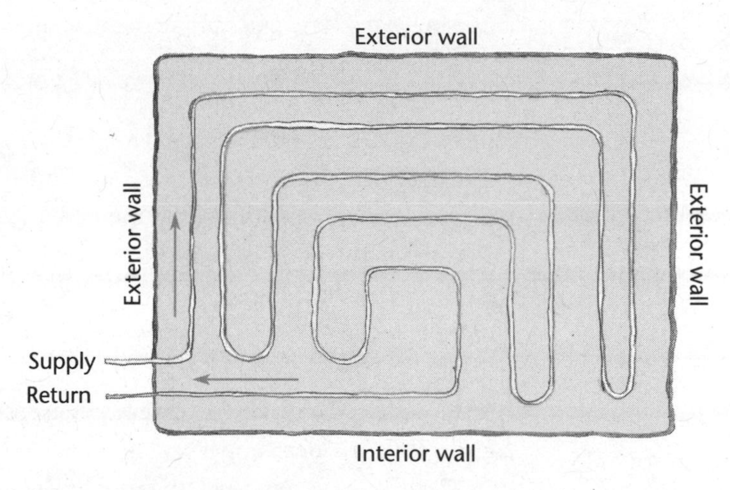

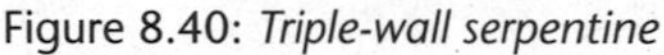
Figure 8.40: *Triple-wall serpentine*

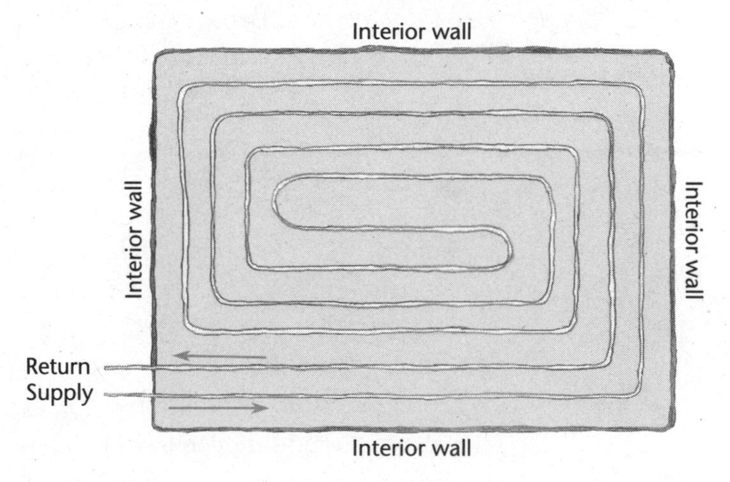

Figure 8.41: *Counterflow spiral pattern*

Pex Tubing

The first modern radiant floor heating systems used copper or steel pipes, which were located beneath the floor, usually in concrete. Hot water was circulated through these pipes. These systems were problematic because the pipes tended to break, causing leaks. The pipes broke because the metal expanded and contracted throughout the heating and cooling cycles. The pipes expanded and contracted more than did the surrounding concrete, resulting in great stress being placed on the piping. Eventually the pipes broke from the stress. Another problem arose when the concrete cracked. The metal pipes were not flexible, so they tended to crack along with the concrete. To make matters even worse, concrete corrodes copper.

Pex tubing is technically a cross-linked polyethylene. This product was developed in Northern Europe specifically to meet the special demands associated with radiant floor heating. Heating costs in those areas are double or triple what they are in the USA, and radiant floor heating systems were known to use the least amount of fuel to heat a building. So a reliable tube was needed to do this job, and Pex tubing has stood the test of time. Pex tubing is flexible but very strong. It is nearly inert and does not corrode. It can also withstand moderately hot temperatures without deteriorating. It is a nearly perfect product for use in this type of solar heating system. The advent of Pex tubing has allowed the advancement of high-mass solar heating systems.

We suggest using a minimum size of ½-inch ID tubing for the Pex tubing loops in the floor or sand bed. Some people use ⅝-inch ID tubing. Both sizes work well. There will be slightly less resistance to flow in the larger tubing. You will need 1 to 1.25 feet of tubing for every square foot of floor area.

Tips for Installing Pex Tubing

Pex tubing comes coiled from the supplier. Though this tubing is flexible, it can be stiff when cold. Always keep the tubing in a warm place for at least six hours before uncoiling. When uncoiling it, let the tubing sit in the sun for a while before placing it in its final location. This will soften up the tubing and make it easier to place.

When removing the tubing from the box, make sure you do not cut it. The tubing cannot be used directly from the box. You must uncoil the whole roll before placing the tubing. If you try to just take the tubing off the coil without unwinding it, it will not lie flat and will create big spirals of tubing that will get tangled up.

Put a piece of duct tape over each end of the tubing to keep dirt from getting into the tube. The glue from any tape will deteriorate the tubing, so later you will have to cut off the ends that were taped. Remember to account for this when locating the ends of the tubing in preparation for attachment to the tubing manifolds. Never use tape to hold the tubing in place.

The simplest method of uncoiling is to have someone hold the end of the tubing that is on the outside of the coil. Then roll the coil out on the ground as if it were a wheel. Most coils are at least 300 feet long, so you must have at least that much open space to work in. While unrolling the coil, you must keep it from getting away from you. If it falls to the side or some parts of the coil begin to hang from the side, a tangle could happen.

Another easy method of uncoiling is to have one person take the end of the tubing that is on the outside of the coil. The second person holds the coil vertically like a tire on a car, with his arms clasped through the middle. The person with the end then walks slowly away from the person with the coil. The coil person must help the coil unroll by flipping the coil to facilitate uncoiling. This takes a little practice, but can go quite fast.

You can also use an uncoiler. This is like a Lazy Susan, a platform that turns horizontally. Place the coil on the Lazy Susan and while one person takes the end of the tubing that is on the outside of the coil and begins to pull, the other person holds the coil on the uncoiler and helps the tubing come off without tangling.

The warmer it is outside, the easier it will be to install the tubing. The warmth makes the tubing more flexible. It is also helpful if it is sunny, for the same reason. Leave the tubing uncoiled and in the sun for an hour to allow it to soften up. A few hours in the sun won't damage the tubing, but prolonged exposure to the sun is not recommended. The tubing should not be exposed to the sun for more than 30 days. Consult the tubing manufacturer for recommendations. Exposure to sun is usually not a problem for the bulk of the tubing because it is covered with dirt or concrete soon after layout. If, by chance, the tubing will be laid out but not covered for some time, cover it with tarps or anything that will keep the sun off. Take care to cover the tubing ends that protrude from the slab if they will be exposed to the sun for any length of time.

There is a limit to the length of tubing used in a circuit. If the tube length is too long, the hot fluid passing through the tube will cool off before it gets to the end of the circuit, and heating will not take place evenly. For ½-inch and ⅝-inch tubing, the maximum length for each circuit is 300 feet. Tubing circuits can be shorter than 300 feet. Many manufacturers package Pex tubing in 300-foot rolls. It is easiest if all circuits are relatively the same length, which makes balancing the system automatic. This is especially true in high-mass systems. The goal of balancing this part of the system is to have the temperature of the solar fluid where it exits the sand bed the same temperature in all the circuits. When the tubing circuits are all the same length, they will naturally all perform the same way. Shorter circuits will lose less heat because they are shorter, so the exiting solar fluid will be hotter. If the circuits are not balanced, uneven heating could occur. If all circuits cannot be the same length, ball valves can be used to adjust the flow of solar fluid through each circuit.

Your system designer should calculate the best distance between tubing for you. If your specifications call for 12-inch spacing, you will need 1 linear foot of tubing for each square foot of area. A 300-foot circuit will require 300 square feet of area. If the specifications call for 9-inch spacing, a 300-foot circuit will require 222 square feet of area. If the total area to be covered by the gridwork of tubing is 900 square feet, and the tube spacing is 1 foot, you need to divide the total area into three equal sections, one for each circuit. Remember that you will have less than the full length of tubing to work with in most sections because some of the tubing length will be used up in going to and from the manifold location to the area to be covered with the gridwork. Be sure to space these tubing runs the same as all the rest of the tubing.

Once you get the total area divided into sections, it is time to start laying tubing. The surface where the tubing will lie should be prepared properly. The ground should be leveled at the proper level. Insulation is usually laid first. A vapor barrier is also usually incorporated below

Midwest Renewable Energy Association

Figure 8.42: *Radiant floor tubing attached to rebar in concrete slab*

the insulation. If the tubing is to be located in a slab, the total area on top of the insulation can be covered with reinforcing wire or rebar.

Reinforcing wire comes in big rolls or 4' × 8' pieces. The wires usually form 6-inch squares. This wire spacing can help with the tubing spacing. You can follow the wire to achieve a consistent tubing pattern. If using rebar, place the rebar to assist tubing placement. You can tie the tubing to the wire or rebar with nylon zipties or with short pieces of thin wire. Some manufacturers can provide either wires or zipties. Use a fastener every 12 to 18 inches, closer at bends. The tube needs to be firmly affixed to the wire or rebar to hold it in place when the concrete is poured. There are also staples with barbed ends that will hold the tubing to the insulation. These work well and are an alternative to wire or rebar.

The tubing is full of air when the concrete is poured and it tends to float up into the slab. You want to minimize this and keep the tubing as close to the bottom of the slab as possible. Fasteners are cheap. Use lots.

Start where the tubing will exit the slab. Use a board with ¾-inch or larger holes drilled through it. The holes should be 2 inches apart, and there should be two holes for each circuit drilled in a row. This board will help hold the tube ends in a neat row. Leave plenty of tubing protruding past the slab top to allow for connecting the tubing to the manifolds. Think about where the manifolds will be located and allow plenty of extra tube length. Remember that you will have to trim the ends of the tubing.

Bob Ramlow

Figure 8.43: *Tube exiting the slab*

In order to protect the tubing where it exits the slab, a piece of electrical conduit should be slipped over the tubing and placed so the top of the conduit extends several inches above the top of the finished slab. Three-quarter-inch conduit works with ½-inch tubing. If you are laying the tubing in a slab, use 90-degree elbows for this. The preformed elbow has a gentle sweep and holds the tubing in place as well as protecting it. Slide the elbow along the tube till one leg of the elbow is parallel with the ground and flat on the wire or rebar. The other leg of the elbow will be pointing straight up and should protrude through the top of the soon-to-be-poured slab. If you are placing the tube in a sand bed, use straight pieces of conduit about one foot long. The 90-degree bend the tube needs to make will be well below the slab, so you don't need the preformed elbow.

Start with both ends of one circuit fixed through the board through adjacent holes. It helps to mark each tube end with an arrow indicating the direction of fluid flow, the hot arrow pointing down and the return arrow pointing up. Do not mark on the tube with a marker. Put the arrows on the tape and on the conduit where it will be above the slab top. This will be very important when it comes time to attach the tubing to the manifolds.

Now start to place the tubing for the first circuit. This circuit should be the farthest from the manifold. The feed end of the tube (from the hot manifold) should be placed nearest the exterior of the building, and the return end of the tube should be toward the inside of the building right next to the feed tube. These tubes should run parallel to each other and be spaced apart at the designated distance. When you reach the area to be covered with this circuit, lay out the tubing in the predetermined pattern. Remember to keep the hot tube closest to the outside of the building. Remember that because this circuit is farthest from the manifold location, the tubes have had to run through the area designated for other circuits. Because of this, the tubing will not quite fill the area that was designated for it. That is okay, because when you lay out the rest of the circuits, some of their area will be covered by the far-away circuits as they make their way to and from the manifolds.

When you are done, you will have a pair of tube ends for each circuit, held neatly by the board with holes in it. You can adjust the placement of the board so it comes up near the header location. After the slab is poured and the building is closed in, you will attach the manifolds to tubing and run the required piping.

Constructing the Sand Bed for High-Mass Systems

What we are calling the sand bed is an insulated box filled with sand, which is buried beneath the floor. Exactly how this box is built will depend on how the building it serves will be constructed. Here are some examples. Please note that it will be up to the contractor to decide what will be the best method, depending on site and climate. His or her experience in building in your area should give you the most stable building.

The important factor here is that the exact design of the sand box is flexible and should be fit into the design of the footings and slab of the building.

Consult the illustration of the high-mass system detail in the description of that system to get a good impression of the order of the layers and their layout. (See also Figure 6.6). If a radon mitigation system will be installed in the building, it should be installed beneath the sand bed.

A vapor barrier can be installed either above or below the insulation. Most contractors prefer to have the vapor barrier on the outside of the insulation.

Installing Pex Tubing in the Sand Bed

If you are placing the tubing in a sand bed, you will not have any wire or rebar to attach the tubing to. Place the vapor barrier and insulation as described above. Next, fill the sand bed to the depth of the tubing gridwork and level it. In a sand bed, it is not as critical to achieve a per-

Figure 8.44: *Insulated box for high-mass sand bed*

MARGUERITE RAMLOW

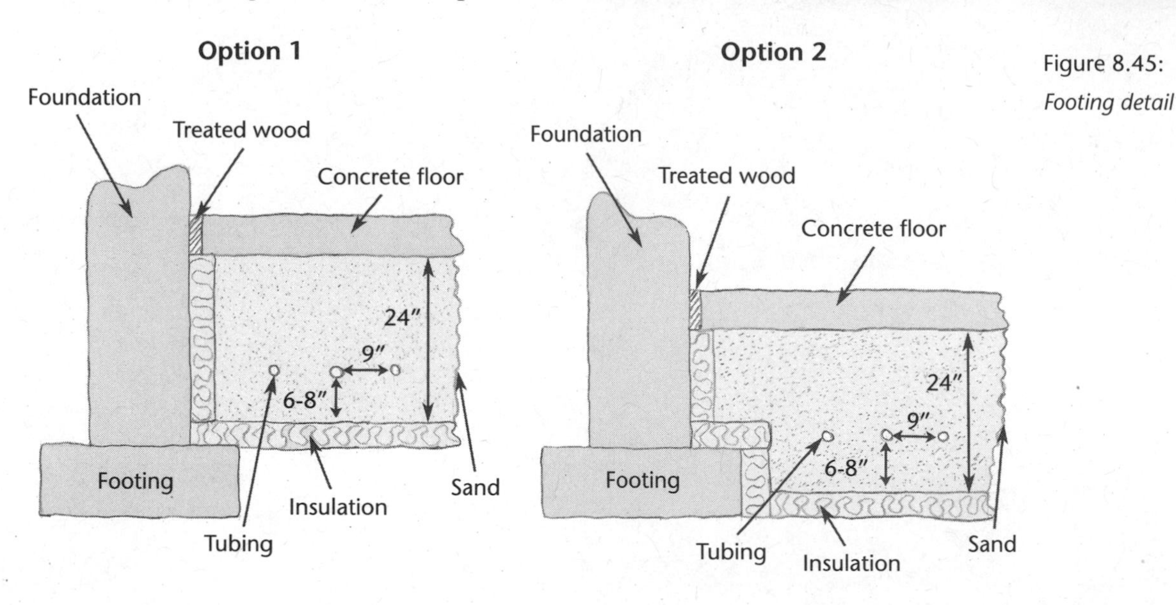

Figure 8.45: *Footing detail*

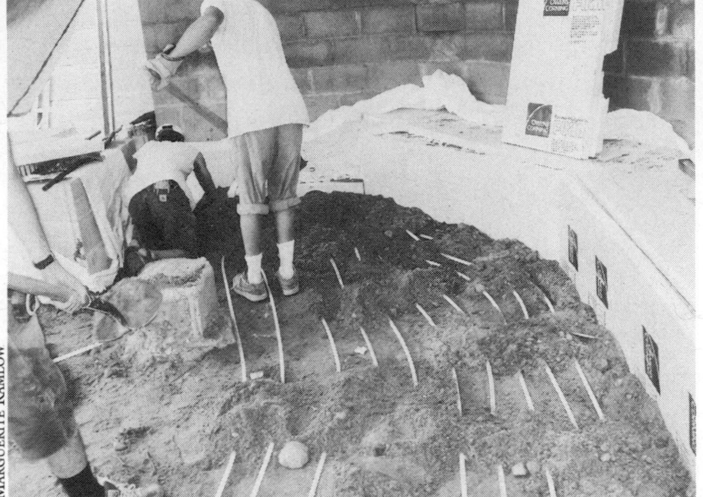

MARGUERITE RAMLOW

Figure 8.46: *Installing tubing in a high-mass sand bed*

fect grid of tubing (the sand is somewhat forgiving compared to a slab), but you still want to have a fairly even distribution of tubing. As you start to lay the tubing, you can hold it in place with sand. It is helpful to have one person placing the tubing and another with a shovel plopping piles of sand on the tubing once it is in place. If the sand is damp, you will only need a shovelful every several feet. With dry sand, you may need a continuous covering.

Wherever the tubing makes a bend, plenty of sand is needed to hold it in place. It takes some practice not to pull the tubing once it is laid. Even in warm conditions, the tubing seems to have a mind of its own and will want to curl up on you. Once you develop a successful method, it goes fast. Compact each layer of sand as it is added. If you have trouble keeping the bends in place, some contractors lay a piece of rebar down and attach the tubing to the rebar.

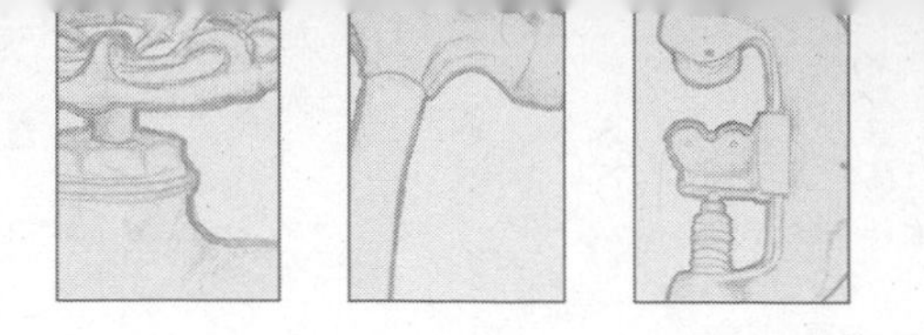

9

SYSTEM OPERATION AND MAINTENANCE

Whether you are a professional installing a system for a client or if you are installing a system for yourself, please provide an owner's manual for the system. This manual should include a line drawing of the system, a verbal description of the system and how it works, and warranty and owner manuals for all components used in the system. The owner's manual should be kept at the system and not in a file cabinet somewhere. We suggest putting the manual in a plastic sleeve and attaching the sleeve to the solar storage tank. The system should also be properly labeled, with the insulated piping labeled as to direction and whether it has solar fluid or water in it. There should also be a tag on the heat exchanger stating the brand of solar fluid, the dilution rate and the date of installation. Including this information at the end of the installation is critical to prolonging the life of the system.

Charging a Pressurized System

Charging a system is the process of adding the solar fluid to the closed loop and getting all the air out of that loop.

CAUTION!

All liquid-type solar collectors can produce steam! Extreme caution is advised whenever filling or draining closed-loop solar energy systems. Always cover solar panels when filling or draining a solar energy system if the sun is shining on the collectors. Never open a closed-loop system when the panels are exposed to the sun. This is no joke! Be careful. Failure to follow the above precautions could result in severe burns.

Please read through this whole procedure before attempting to charge your system.

List of materials needed to complete the task:

1. Charging pump
2. Three 4-foot hoses with female hose fittings on both ends of one and on one end of the other two
3. Nontoxic glycol to fill the system to the proper temperature rating
4. Distilled water to dilute glycol
5. Channel-lock pliers
6. Rags
7. Empty 5-gallon pail
8. Gloves
9. Trouble light

The charging pump should be a booster type, large enough to pump the solar fluid to the top of the collector array. The greater the vertical distance between the charging ports and the top of the collector array, the larger the size of charging pump that is needed. We use a one-half-horsepower jet pump, which can pump more than 40 vertical feet with no problem. Many plumbers use this type of pump regularly to fill hydronic heating systems. They are also relatively inexpensive to purchase. The pump needs to be outfitted with male hose fittings on both its inlet and outlet. Male hose fittings can be purchased at any good hardware store. You will need a male-hose-to-male-NPT thread adapter and perhaps a hex bushing to attach to the pump.

Professionals have a dedicated charging pump. They usually outfit it with additional components to make the charging job easier. Figure 9.1 shows a common configuration. On the outlet side of the pump, the piping is typically copper and is fitted with a pressure-relief valve, pressure gauge, ball valve and check valve.

This pump will first have to fill the system with solar fluid by sucking premixed solar fluid from a five-gallon pail. Once the system is full, the pump will then vigorously circulate solar fluid throughout the system, thereby flushing out all the microscopic air bubbles that can be difficult to purge any other way. Finally, the pump will be used to pressurize the system to the appropriate pressure for your location and design.

Filling the System

Locate the charging ports in the plumbing header. The charging ports are the two boiler drains with the check valve or ball valve between them. Set up the charging pump near enough to the charging ports so that each length of hose can easily reach between. Set the empty five-gallon

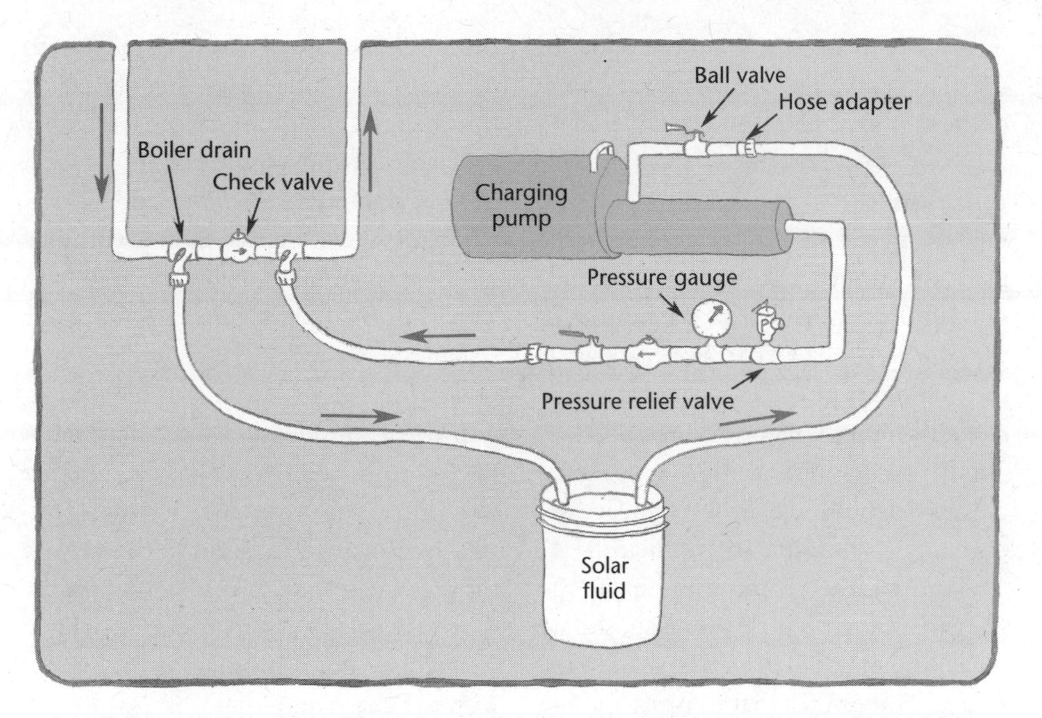

Figure 9.1: *Charging pump configuration*

pail near the pump and charging ports. Attach one end of the hose with two male ends to the outlet of the charging pump and attach the other end to the charging port that the arrow on the check valve is pointing to. This is the input valve. Fully open the input valve. Attach the second hose to the other charging valve (the drain valve) and put the other end of that hose into the pail. The end of this hose must reach the bottom of the pail. Fully open the charging port drain valve. Attach the last hose to the inlet of the charging pump and put the other end into the pail. This hose also must reach the bottom of the pail. See Figure 9.1.

If there are any ball valves in your system, make sure they are all fully open.

Fill the pail with premixed glycol and distilled water. (See the section on solar fluids in Chapter 4 for the proper dilutions.) You are now ready to begin charging the system with fluid. You also must be able to see the pressure gauge at all times while charging the system. If you cannot see the pressure gauge, have a helper stationed at the pressure gauge to give you readings at all times during the charging process.

If you have a self-priming pump, you are ready to start. If you do not have a self-priming pump, please prime it now by removing the hose from the outlet of the pump and filling the pump with some of the premixed glycol and then replacing the hose. Firmly tighten all hoses with your channel-lock pliers. With one hand, grab the two hoses that are going into the pail. You want to make sure that the ends of both hoses remain near the bottom of the pail. Never let go of these hoses while the pump is running, and never leave the charging pump running unattended. A popular technique is to use spring loaded clamps to hold the hoses in place by clamping them to the top of the pail.

Turn on the charging pump. If the pump has been properly primed, it should start pumping fluid from the pail into the system. You can tell that things are going properly if the fluid level in the pail begins to go down. Make sure to monitor the hoses in the pail so they don't start flopping around. Carefully watch the pressure gauge. You want the pressure to stay between 10 and 25 pounds per square inch. You can control the pressure by adjusting the inlet or outlet filler valve. You can use the valves to adjust the pressure to maintain the correct pressure.

Right after you turn on the pump, air will start to come out of the system through the drain hose that is in the pail. Because the end of this hose is at the bottom of the pail (below the fluid level), bubbles will begin to rise in the pail. As the level of fluid in the pail begins to drop, add more, trying to keep the pail at least half full at all times. Keep the empty pails you accumulate nearby in case you need to drain fluid from the system.

While the system is filling, you may hear various gurgling and bubbling sounds. This is normal. At a certain point, your system will be nearly full of fluid and devoid of air. You can tell because the bubbling in the pail subsides and the fluid level in the pail stops going down. Make sure that you are monitoring the pressure constantly. When the system becomes full, the pressure will want to go up. Adjust the valve to maintain proper pressure. Continue to circulate fluid throughout the system with the charging pump.

If you find that, as you begin to pump fluid into the system, the pressure rises quickly and keeps going up, while at the same time no air comes out of the drain port and hose, you should turn off the pump before the pressure gets above 50 pounds per square inch. Be careful at this point, because pressurized fluid may flow back vigorously into the pail through the pump. If fluid will not circulate into the solar loop but the pressure builds quickly, your system may be plugged or it may have an air lock. First, check that all ball valves within the system are open. Next double check that the plumbing runs have been hooked up properly. You should be able to follow the pipe runs from the inlet port up and through the collectors and then back down through the heat exchanger and finally to the outlet port. If you find that all valves are open and the pipe runs are plumbed properly, then you may have an air lock in the system.

A good way to overcome an air lock is to close the inlet port valve while leaving the drain valve fully open. Turn on the pump and slowly crack open the inlet valve. The pressure in the system will slowly rise. Allow the pressure to go to 60 pounds per square inch, and then close the filler valve and shut off the pump. Make sure the drain valve remains fully open, and always hold the drain hose firmly. Usually this pressure is enough to push any air bubbles through the system after a few moments and to allow the solar fluid to continue through the system and back to the drain port. If there is still no circulation, try closing and opening all ball valves. Sometimes air bubbles can hang up in these valves. As a last resort, bring the pressure up to 75 pounds per square inch. The pressure-relief valve may begin to open at this point. Never bring the pressure above 75 pounds per square inch, and always hang on to the drain hose firmly.

If you have an assistant, have him or her look over all the piping to check for leaks. Make sure to check the fittings between each collector (if you have more than one) and wherever there is a threaded fitting in the system. If you did everything right, there will be no leaks. If you do not have an assistant, you need to check it out yourself. To do it yourself, you must put the system under some pressure, and you must turn off the pump. To do this, simultaneously close both filler and drain valves on the system. Watch your pressure! When both valves are closed, check the pressure. If it is above 23 pounds per square inch, turn off the pump and go check the system for leaks. If it is under 20 pounds per square inch, then slightly open the filler valve only. Watch your pressure! When the pressure reaches 25 to 40 pounds per square inch, close the valve and turn off the pump. Go check your system.

If you discover a leak, you will have to drain the fluid from the system and go back and solder the leaky joint or retighten the fitting. If you found no leaks, continue with the charging process.

The next step is to continue to flush all air from the closed loop and, lastly, to pressurize the system. If you had to shut off the pump, turn it back on, and then slowly open both valves, opening the drain valve first. As always, watch your pressure to maintain 10 to 25 pounds per square inch pressure in the system. The pail must be more than half full at this point, and you will need a good light to observe the fluid in the pail. You should observe the pressure in the system remaining constant, and you will see a swirling action within the pail created from a fast stream of solar fluid coming from the drain hose. The end of the drain hose has always remained near the bottom of the pail. You will notice that the color of the solar fluid is milky white, and that surges of air will often bubble out. You must observe carefully what is going on in the pail. Closely observe color changes and bubble size. You have to look carefully during the final stages of purging because near the end of the procedure the bubbles are very small and hard to see. Nonetheless, you need to get as many of those little bubbles out of the closed loop as possible.

After the pump has run for a while, the amount of bubbling will begin to subside and the milky color will fade. The milky color is caused by millions of microscopic air bubbles. The amount of time it takes to get rid of all the air depends on the speed of the charging pump and the total length of the piping in the system. It is sometimes helpful to periodically turn off the charging pump to let the bubbles settle out of the charging pail and then resume the charging process. Additional help to get all the air out of the system is to periodically let the pressure rise in the system and then quickly release the pressure. You can accomplish this by kinking the drain hose while carefully monitoring the system pressure. You can also kink and release the hose quickly, which causes the piping to shake and release bubbles that may be lodged within the system.

Pressurizing the System

We recommend that you continue to circulate fluid for at least five minutes after the last bubbles are observed in the pail. Then slowly begin to close both filler and drain valves, maintaining at least ten pounds of pressure per square inch. When both valves are closed, slowly crack open the filler valve to allow the pressure to rise to at least 35 pounds per square inch. This is most easily accomplished by allowing the pressure to rise to above 35 pounds per square inch before shutting off the filler valve. The pressure gauge may bounce around a little while you are opening and closing valves. Turn off the pump. Let the pressure settle down. Then slowly crack the drain valve to release pressure down to at a minimum of 35 pounds per square inch at 60°F solar fluid temperature. If the solar fluid is cold, lower the charging pressure to 30 pounds per square inch.

You are almost done. Have your rags handy and grab the channel-lock pliers. Tighten both charging valves very tightly. Remove the hoses from the charging valves. Install caps on the charging valves and tighten them securely. As you recall from previous chapters, liquids expand and contract with temperature changes. This will be reflected in pressure changes on your system gauge. At noon on a clear day in the middle of summer, you may observe pressure of more than 40 pounds per square inch, but on a night when it is 20°F below zero you may observe 15 pounds per square inch. But most of the time, you should observe pressure readings of about 30 pounds per square inch. If you notice a consistent drop in pressure, you have a slow leak. Thoroughly check the system, paying special attention to nonsoldered fittings. If you are having trouble locating a leak, try isolating individual circuits and observing pressure changes. After you've determined that the system is operating properly and with no leaks, install caps on the filler/drain ports as a safeguard against accidental opening. Place an informational label on the expansion tank stating date of installation, servicing dealer or plumber, type of solar fluid, solar fluid concentration and initial pressure. Clean up and you're done.

Charging a High-Mass System

If you have a combination system with multiple load circuits, you now need to flush each individual circuit. During this procedure, always keep at least one circuit open at all times.

If you are installing a system with a bypass pipe around the hot water heat exchanger, it should have a ball valve installed on the bypass. Close that valve halfway. You should have a main heating circuit and a shunt circuit. Completely close the shunt loop valve, diverting all the flow through the heating system. If you have valves on each individual tubing circuit (recommended), leave two circuits open and close the rest. If your tubing is translucent, you can observe bubbles being flushed from the system. After those loops are clear of bubbles, open the next two circuits and then close the ones that were flushed. Work through the system till all are flushed. After the floor circuits are flushed, open the shunt loop valve and then close the floor valve. Pump till no more bubbles are observed. Finally, open all circuits again.

Finishing Up

After you are done charging the system, you should clearly label all the main lines with identification tags. The tags should identify what the lines are and the direction of flow. You should also assemble a booklet about the system that contains the following information:

- operating instructions
- collector literature and warranty
- storage tank and expansion tank information and warranty

- pump information and warranty
- description of system operation including a line drawing
- controller information and warranty (if applicable)
- solar fluid information
- maintenance information

Controls and Power Sources

All solar water and space heating systems, except for thermosiphon and ICS systems, require a pump to circulate solar fluid through the solar loop. A circulation pump has to operate whenever there is solar energy to harvest, but it cannot run during noncollection hours. If the pump ran all the time, the solar collection system would become a heat-dissipation system, especially at night when it is cold. Therefore, if the pump can't run all the time, there has to be some kind of brain to turn it on at the appropriate times as well as turn it off when there is no longer solar energy to collect or when the storage is up to its maximum temperature.

There are basically two ways that a circulating pump is controlled. The traditional method is to use an AC-powered pump controlled by a differential temperature controller. Basic information about controller operation was introduced in Chapter 4. The other method is to use a DC-powered pump that runs directly off a photovoltaic module.

AC-Powered Controls

In AC-powered systems, the power to run the pump comes from our regular 120-volt AC home electrical system using utility-generated electricity. Of course, that electricity is "on" 24 hours a day, so if the pump is plugged into an outlet, it would run all the time, which you don't want. A differential temperature controller uses sensors that electronically measure temperatures at remote locations. In a solar water heating system, one sensor is placed to measure the average temperature of the storage, while another sensor is placed to measure the temperature of the collectors. The differential temperature controller compares the temperature of the storage and the temperature of the collectors. Whenever the collectors are hotter than the storage, the controller switches the pump on. The pump will continue to run till the temperature in the collectors nears the temperature in the storage.

A modern differential temperature controller uses a small microprocessor to analyze the information it gets from the sensors, and a logic chip is used to decide when to do the switching. The controller is powered by 120-volt AC and is either hardwired to a fused circuit in the building's AC load center or is plugged into an outlet like any other household appliance. The microprocessor operates on low voltage, so there is a transformer in most controllers. Note that the sensors also operate on this low voltage. In these systems, the pump is 120-volt AC powered. To turn the pump on and off, a relay switch is also included in the controller.

Numerous brands of differential temperature controllers are available on the market. Most of them are designed to work for various types of solar water heating systems. Usually a number of settings must be adjusted when the controller is installed so it will function properly for your particular system. Many also include a high-limit function that turns off the pump when the storage reaches a preset temperature. This is to prevent storage overheating. This is a fine way of protecting the system if the solar water heating system is a drainback system. The problem with this approach with pressurized systems is that in a pressurized system we want circulation whenever the sun is shining. Some newer controllers have a second relay that engages when the storage reaches high-limit while the main relay stays on. The second relay is used to divert the solar fluid to a diversion (shunt) load by engaging a valve or pump.

Many controllers also have a recirculation mode, which is used for flooded systems. When the recirculation mode is enabled, the controller will turn the pump on when the collector approaches freezing conditions. Warm water is then drawn out of the storage tank and circulated throughout the system, offering protection during occasional freezing episodes. When shopping for a differential controller, make sure it has the features you will need for the type of system you are installing. Also make sure to read the instructions that come with the controller, so you get the settings correct.

In the 1980s, many companies made their own controllers, which were often very simple devices. Today, most AC-powered systems use mass-produced controllers. Although today's controllers are very reliable and affordable, replacing controllers has been one of the most common service calls over the years. The problem is that this type of electronic device is vulnerable to electrical discharges, caused primarily by lightning but also by spikes from the electric utility. Most of the failures were due to improper grounding of the solar energy system. If you suspect lightning damage, open the cover and look at the circuit board for burn marks or black blotches. Sometimes the mark is hard to find, but often it is big and obvious. Relays are sometimes burnt as well. The relay is usually enclosed in a plastic cover, which is often clear.

A few controllers are made specifically for DC or PV direct systems. These controllers address some of the criticisms of PV direct systems by offering differential

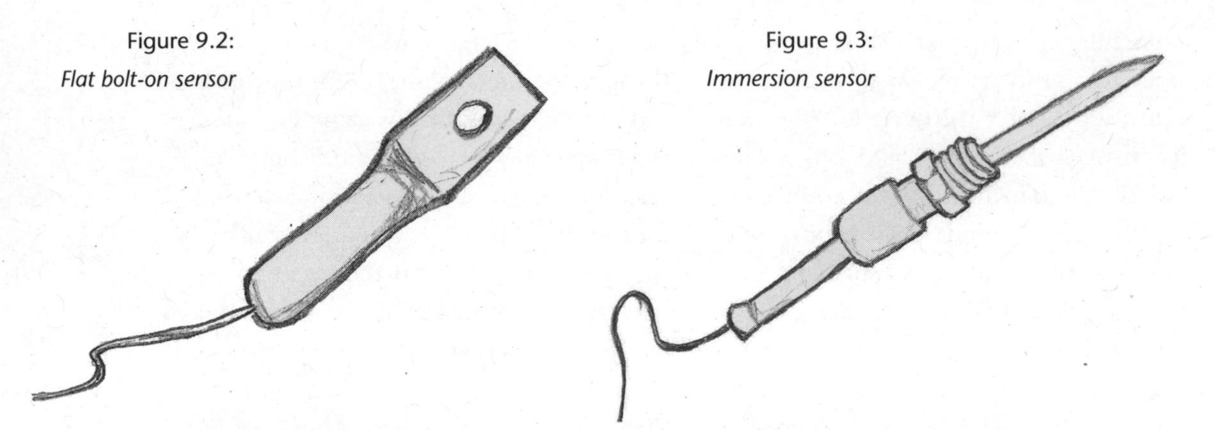

Figure 9.2: *Flat bolt-on sensor*

Figure 9.3: *Immersion sensor*

temperature functions. These controllers also offer the attractive option of having digital readouts for the collector and storage temperatures.

To properly install a differential temperature controller you must properly ground your whole solar water heating system. All the collector frames must be bonded together with a copper ground wire and proper fasteners.

Sensors are typically ¼-inch-diameter by 1-inch copper cylinders that have two wires coming out of one end. Sometimes one end is flattened. Two wires connect the sensor to the controller. Every controller requires a specific type of sensor to operate properly. Most controllers use sensors that are identified as 10K sensors. These can be identified by their black wires. Some older controllers use 5K sensors, which are identified by white wires. It is best practice that you use # 16 stranded sensor wire that is shielded. The shield should be connected to the ground lug in the controller, and the other end of the shield should be left unattached to anything. When wiring the sensors, it is best to solder all electrical connections or use watertight connectors such as waterproof telecom splicing connectors. The connections then should be sealed with heat-shrinkable tubing. Using wire nuts allows corrosion to form at the connection, causing inaccurate sensor readings. This is especially important for the collector sensor, which is exposed to a harsher environment than the tank sensor.

The sensor that is located at the collector is usually mounted on the pipe that exits the collector and delivers the hot solar fluid to the heat exchanger. The sensor should be located as close to the collector as possible. You may be able to actually slip a part of the sensor under the rubber gasket where the pipe exits the collector frame. The sensor is attached to

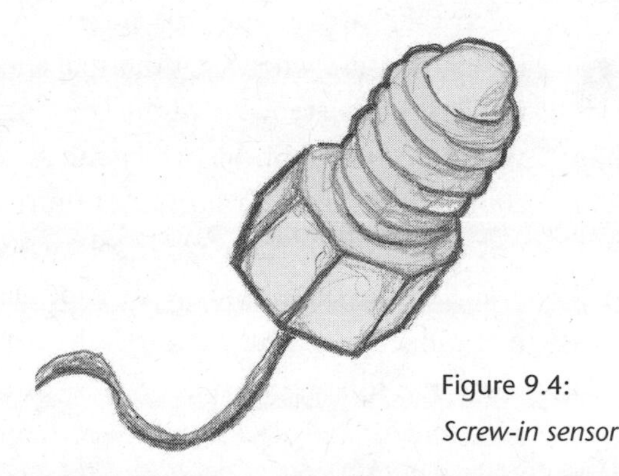

Figure 9.4:
Screw-in sensor

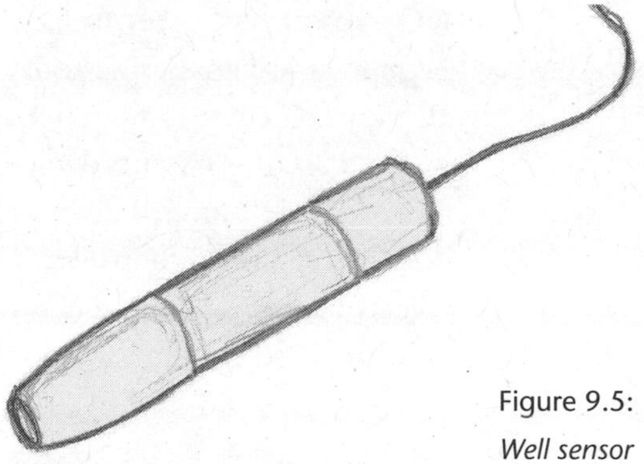

Figure 9.5:
Well sensor

this pipe with a stainless steel hose clamp. Some installers locate the sensor on the top pipe between the collectors (if multiple collectors are used), using a stainless steel hose clamp. Either location is acceptable. The best location for this sensor is actually inside the collector, with the sensor attached to the top of the top manifold. This is particularly useful if the system is a drainback design to ensure optimal system operation. Some collector manufacturers offer this option, but make sure the sensor they install is compatible with your controller. You can also install a sensor at this location by removing the glass from the collector in the shop and installing the sensor. Route the sensor wire out the back of the collector by installing a grommet in the back plate of the collector.

The tank sensor is typically mounted about one-quarter to one-third of the way up from the bottom of the storage tank. The sensor must make good contact with the side of the tank. Most solar storage tanks are insulated with foam, so the sensor must be mounted between the foam and the tank. Most tanks have an access door near the bottom of the tank. Some tanks have a stud welded to the side of the tank, allowing easy sensor mounting. If no stud is available, the sensor needs to be firmly pressed to the side of the tank. Often, a piece of foam insulation can be used to wedge between the existing insulation and the side of the tank jacket to hold the sensor firmly against the tank. This is a common practice when using an electric water heater as the storage tank. Another option for tank sensor location, if you are using an electric water heater as the storage tank, is to remove the lower heating element and use a screw-in or well type sensor in that port. This is especially good if the storage tank is fiberglass or plastic.

DC-Powered Systems

A popular system design today utilizes a 12-volt DC circulating pump and a dedicated 12-volt DC photovoltaic module to power the pump. This type of system does not require a controller. The photovoltaic module is mounted near the thermal collectors, so the only time it produces electrical power is when the sun shines on it, which is also when solar thermal energy is being collected. At night, or during very cloudy conditions when the system should not be operating, the pump will not run because there is no solar energy available for the photovoltaic module to make electricity. During low-light conditions, the solar thermal collector's output is reduced and so is the photovoltaic module's. If there is enough solar energy available to create heat in the thermal collector, there is enough to make electricity in the photovoltaic module as well, but at a reduced rate. The pump will vary its speed in direct relation to the amount of solar energy available. This is a perfect match, as the pump speed exactly matches the circulation needs of the solar thermal collector.

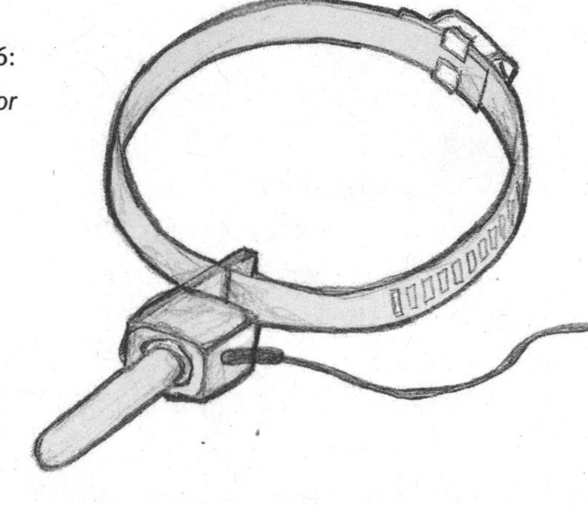

Figure 9.6: *Pool sensor*

Wire Nuts

Wire nuts are used to fasten two wires together and are commonly used in 120-volt AC electrical wiring. Wire nuts should not be used in low-voltage wiring, especially on sensor wiring circuits. A sensor is a resistor. It has different resistance for each temperature. Corrosion on a wire is also resistance that may give false readings to the differential temperature controller. For these low-voltage applications always use waterproof connectors or solder the connections and use heat shrink tape to seal the connection. This is especially important when the sensor wire connection is exposed to the elements.

In climates that experience significant snow, it is best to locate the PV panel in a place where it can be seasonally adjustable. During the snow season, the PV should be adjusted to be vertical to eliminate snow buildup that can stop system operation, because even a little shading on a PV panel can totally turn it off. During the rest of the year, the PV panel should be adjusted to the same angle as the thermal collectors.

When choosing a 12-volt DC circulating pump, it is important to pick one that is designed to operate directly off a

Bob Ramlow

Figure 9.7: *Exterior pump enclosed in weatherproof box*

photovoltaic module. Brushless pumps are best. Most brush-type pumps will also work well, but the addition of a linear current booster in the circuit between the PV panel and the pump often increases the reliability of the circulating pump, as the booster will help the pump start properly.

With any electrical circuit, wire sizing is important. With low-voltage circuits, it is important to use the proper wire size for proper system operation.

There is a direct relation among system voltage, amp draw of the circuit and length of the wire. In general, the lower the voltage of a particular circuit, the larger the wire size needed. If the wire is too small, not enough current will flow through, and in this case the pump will not operate properly. Please refer to the wire-sizing chart in Appendix A when sizing your wire from the photovoltaic module to the pump. The figures you will need when referring to the chart are the amp draw of your pump and the length of wire needed to connect the photovoltaic module to the pump.

The traditional location for the main circulating pump is near the heat exchanger and storage tank. This is still the case with AC-powered systems. With low-voltage systems like photovoltaic-powered systems, it is common to locate the main circulating pump near the photovoltaic module and the thermal collectors. The main reason for this is to save money and resources by eliminating a long wire run. This wire run can get expensive because of the 12-volt DC circuit requiring large wire sizes. If the collectors are roof mounted, the pump can be located either in an accessible attic or near or behind the collectors in a ventilated enclosure to protect the pump from the elements. If the collectors are ground mounted, the pump can be mounted behind the collectors in a weatherproof and ventilated box. It is important that this box is ventilated to prevent overheating.

Grounding Collector Arrays

It is important to properly ground all solar thermal systems, but it is particularly important to ground AC-powered systems to protect the controller. Solar collectors can act like big static electricity absorbers, so when lightning strikes nearby, some of that electrical energy is caught by the collectors. The whole system is tied together by copper pipe, so that static can be routed through the system to the controller.

When installing collectors, it is important to electrically bond all the collectors together and to the copper piping and then to a dedicated ground rod. A typical grounding system uses #6 bare and stranded copper wire, which is screwed to each collector and then bonded to the system piping; it is then connected to the common ground rod. When attaching the stranded ground wire to the collectors, you can spread out the strands, place a stainless screw through the wire and screw the wire tight against the collector. If the collector is painted, it is best to scrape off the paint where the ground wire is attached to the collector. The best grounding systems always use a continuous grounding wire. Do not splice the wire; use one long strand. Start at the collector furthest away from the ground rod; attach the wire to each panel as the wire travels to the other end of the array; attach the wire to the piping system with a stainless hose clamp; and then run the wire to the ground rod that is used for the building. Note that you should not add an additional ground rod but use the common ground rod that is already serving the building.

AC And DC Systems Compared

Some types of solar water heating systems work only with AC-powered pumps. Draindown and drainback systems require high-head pumps, which are not available in 12-volt DC at this time, so they are always AC-powered and use a controller. Any system that requires any type of motorized valve will also have to be AC-powered, but these are rare.

With AC-powered systems, the circulating pump runs at a constant speed because the power source remains constant. With photovoltaic-powered systems, the pump speed varies depending on the intensity of the sun. This difference in pump operating speed affects the overall performance of the system in a marginal way.

With AC-powered systems, when the pump starts in the morning it is running at full speed while the amount of available solar heat is minimal. If there is still some heat in the storage tank, or if it is very cold outside, the tank can be cooled slightly till the sun's intensity becomes greater. This same situation can also happen during partly cloudy weather as clouds pass over the collectors. This could result in some inefficiency in overall performance. On the other hand, in AC-powered systems the system will turn off when the storage tank either reaches its upper limit or when the tank is warmer than the collectors. This is a good thing in most cases. The only time this is not good is with pressurized systems. Under high-heat situations, if the system isn't circulating, the solar fluid could get extremely hot, which could lead to deterioration. This would happen under low-demand situations, such as when the household is on vacation during the summer. As mentioned earlier, some controllers offer heat diversion options, and we highly recommend this type of controller for pressurized systems.

With DC-powered systems, when the pump starts in the morning it is running slowly because the solar resource is low at that time. The result is that the pump speed exactly matches the output of the solar thermal collectors, resulting in optimum performance. This same situation can be observed under partly cloudy conditions, where the pump speed will always match the output of the solar thermal collectors.

On the other side of the coin, at the end of the day, when the solar thermal collectors are cooling off and the storage tank is hot, the pump is still circulating, which could result in a slight cooling of the storage tank. But because the solar fluid is circulating slowly at this time, the solar fluid stays in the collector longer and actually gets hot. During low-demand conditions mentioned above, the photovoltaic-powered system will continue to circulate the solar fluid, which helps extend the fluid's life by keeping it from overheating.

In the end, the advantages and disadvantages of each system, from a thermal performance perspective, are a wash. The most reliable type of system is a photovoltaic-powered system because it does not require a controller, which is probably the most vulnerable component in a solar water heating system.

Adjusting a Solar Radiant Floor Heating System

In most cases, only seasonal adjustments are necessary on your solar radiant floor heating system. You have a ball valve located on the liquid-to-liquid heat exchanger bypass. This valve should remain about one-third closed all the time. This valve forces some of the hot solar fluid coming down from the collectors through the heat exchanger while allowing most of the fluid to go directly to the radiant floor or to the shunt loop, depending on the season. You need enough flow through the heat exchanger to get enough hot water, while still getting a significant flow to the heating system. Observe the temperature of the tank after a sunny day. If it is anything but exactly 140°F, then adjust the valve accordingly. This valve is adjusted when the system is first set up and will not need further adjustment.

In the spring, when no more heating is required in the home, open the shunt loop valve and close the valve to the radiant floor. Reverse this procedure in late summer or early fall, depending on your heating requirements. You would like to start charging the floor as early as possible in the middle part of the summer. You can adjust the temperature in the home by keeping windows open. Each individual home will react differently, and no two years are exactly alike regarding weather patterns. Of course you don't want to cook yourself out of the house during late summer, so experiment. Bob's high-mass system has never been turned off since it was installed several years ago and it has never been too hot inside the home.

Maintenance

Like any other mechanical devices or products, solar energy systems require periodic maintenance. The good news is that a properly designed and installed system will require very little. You can further ensure minimum maintenance by choosing high-quality parts when installing your system in the first place. We have seen people choose inferior products because they are less expensive, while others choose only high-quality components. The overall price difference between these two systems may be only several hundred dollars, which is insignificant to the overall system cost. Over the long run, the system ends up costing a lot more because of service calls on those inferior components' failures. Save yourself money and headaches down the road by choosing high-quality products in the first place.

Periodic Inspection

It is wise to periodically look over your solar energy system to make sure it is working and that there are no glaring problems.

Once a month you should look at your collector array and observe if there is anything out of the ordinary with it. Glance at the collectors to make sure they are in one piece. Notice any discoloration or dripping. Look for any loose pipe insulation or hanging wires. You can also check once a month to make sure the system is collecting heat. After a sunny day you should check the solar storage tank. If it is hot, all is well. This should take about two minutes a month. Note that it is normal to see condensation on the inside of the glass of flat plate collectors of all kinds. Flat plate collectors must breathe, and they blow out air when they get hot and suck in air as they cool. If the air they suck in has a lot of moisture in it, that moisture ends up condensing on the inside of the glass. This amount of water inside a collector does not deteriorate it.

In addition to periodic visual inspection of your system, some periodic maintenance may be required that is specific to your type of system. We have itemized the main topics below. As there are only three kinds of solar water heating systems that we recommend, we are detailing issues for only those systems.

Differential Temperature Controllers

Because differential temperature controllers are used in several types of systems, we have included a special section just about troubleshooting them. To perform some of the tests listed below, you will need a digital multimeter.

If your system is not operating, the first thing to check is the controller. All controllers will have at least two lights on the front, one indicating that power is being supplied to the unit and the other indicating whether the pump is turned on or off. If both lights are on, take your multimeter and see if power is going to the pumps. If there is power going to the pumps, your problem is with the pumps. If both lights are off, first check that there is power at the unit by using a multimeter. If there is power at the controller and no lights are on, the controller has failed and should be replaced.

Sometimes the power indicator light will be on but the controller is not turning the pumps on at the proper times. This could be caused by a sensor failure, corrosion at sensor wire connections or a controller failure. Visually inspect the circuit board of the controller. If there are any black blotches on the circuit board, it is probably shot. This is often observed after a lightning storm. If there are no obvious burn marks on the circuit board, you can test the controller by removing the sensor wires one pair at a time. With most controllers you can short circuit one of the sensor terminals with a jumper wire, which makes the controller think that the collector is hot. The pumps should come on. If the pumps come on when you short circuit these terminals, there is a problem with the collector sensor or the sensor wire connections. Depending on the type of sensors you have, you will be connecting the jumper wire to either the collector or storage sensor terminals. Remember that either a sensor or a jumper must be attached to both pairs of sensor terminals in order to perform this test.

If the controller checks out, the next thing to check is the sensors. Inspect all connections on the sensor wire circuits for corrosion. These connections should be soldered. If these connections look good, you should then check the sensors. To check a sensor, disconnect the sensor wires from the controller and attach the sensor wires to your digital multimeter and set the meter to measure ohms resistance. If you get no resistance or infinite resistance, the sensor has failed and needs replacing. Most controllers have a sensor resistance chart included with the instructions, so if you have a reading, you can actually tell the temperature of the sensor by comparing the meter reading with the chart. If the reading gives a temperature that is obviously wrong (such as the sensor saying it is 500°F), the sensor is probably bad and needs replacing. If you do not have a sensor chart, a 10K sensor will have a resistance of 10,000 ohms at 77°F and a resistance of 2,043 ohms at 150°F. You can see that as the temperature rises, the resistance goes down. If the sensors check out and the controller still does not do its thing, replace the controller. We have included a 10K resistance/temperature table in the appendix.

ICS Systems

If you live in a climate that never experiences freezing conditions, your ICS system will be in operation all year. The only periodic maintenance required is monthly visual inspections.

If your ICS system is a seasonal system, you will be draining your system at the end of the season and filling it at the start of the season. If the collector is installed in a climate where freezing conditions occur, it is critically important to get the system drained well before freezing conditions are expected. I suggest that you do not push your luck by waiting till the last moment. When draining a system, it is most important to get every bit of water out of the pipes. Even a couple of cupfuls of water can burst a pipe if it freezes. Proper pipe layout and collector orientation during installation will facilitate efficient drainage. Leave all drains open while the system sits idle. If you are unsure whether the system has drained properly, use compressed air to blow out the lines. A few minutes of diligence at this time can save big headaches and money the following spring.

Drainback Systems

Here is what can go wrong with a drainback system: the main pump can fail and the system will simply not work; there could be a loss of solar fluid and the system will simply not work; the controller could fail and the system will simply not work; and the system could freeze and serious complications will arise.

You can detect the first three potential problems by visual inspection. If the pumps fail, you will notice that the system is not heating the solar storage tank. If either pump fails, it fails quickly. Either they work or they don't. If the main circulator pump fails, there will be no heat in either the drainback tank or the solar storage tank after a sunny day. If the heat exchanger pump fails (if one is in the system), then the drainback tank will be hot but the solar storage tank will be cold during operation. If the controller fails, you will notice that neither of the pumps will be running when it is sunny out and the solar storage tank will be cold. Note that if the solar storage tank reaches its high limit, the system may be off even if it is sunny outside, so make sure to check the storage tank for heat if the pumps are not running and it is sunny.

All drainback tanks should be fitted with a sight glass. This sight glass is mounted on the side of the drainback tank and it shows the heat transfer fluid level within the tank. This fluid level should be checked periodically. It is normal for some of this fluid to evaporate over time, so annual close inspection should be done and fluid should be added when required.

System freezes can be caused by a controller failure or climatic conditions. If a controller does not shut the system down properly, the distilled water that is used as the solar fluid could freeze in the collector or in the collector piping. It is unlikely that this would happen. If a freeze occurs, make sure to check the controller for proper operation before putting the system back online.

Pressurized Systems

The wear parts of this type of system are the pumps, the solar fluid and the expansion tank(s). All these components should last between 10 and 20 years before needing replacement. Your monthly inspections should identify whether the system is working. If the system is working, the only thing to do is to verify that the system is holding its pressure. Starting with year ten, the antifreeze should be checked in the fall of the year before serious freezing conditions would be expected. Checking the antifreeze is outlined below. If the antifreeze readings show the fluid is in near-new condition, schedule the next check for five years. If the readings show degradation but still fall within the acceptable range, schedule the next check accordingly. When the antifreeze check shows that the fluid should be replaced, you should replace it. At this time we also like to replace the expansion tank(s).

In a pressurized system, if the system is not circulating and the controller and pumps check out okay and the system is still under pressure, the problem could be caused by a failed expansion tank. Expansion tanks with bladders eventually wear out from the bladders' constant flexing. We always choose an expansion tank that has an extra-heavy-duty bladder. A good bladder-type expansion tank should last 15 to 20 years or more. Bladder expansion tanks should always be mounted below the pipe they are attached to. This helps prolong the life of the bladder by keeping it much cooler than if it were above the pipe it is attached to. You can test a bladder expansion tank by tapping the bottom of the tank. It should sound hollow. If the tank does not sound hollow, release a very small amount of pressure from the Schrader valve on the bottom of the tank. If air comes out, the tank is probably okay. If liquid comes out, the tank is broken and should be replaced.

Checking Antifreeze

The proper antifreeze to be used in these systems is propylene glycol. As this fluid breaks down, it loses its ability to protect the system from freezing, and it becomes more acidic. Special testers are needed to check for these conditions. Regular automotive testers will not work. The most common kind of tester is a refractometer designed to test propylene glycol. You put a drop of antifreeze on the appropriate surface of the refractometer, point the device at a light source, and look into the instrument. You will see a scale that indicates the freeze protection. These meters vary among manufacturers, so follow the directions that come with the tester. Make sure you purchase a tester that measures propylene glycol antifreeze. You can get these meters from some auto supply stores, some solar system manufacturers or scientific instrument companies. As previously described, when propylene glycol/water mixtures get cold, at a certain temperature ice crystals will start to form in the solution and it will start to thicker dramatically. This is called the "freeze point." This is the temperature that most refractometers are calibrated to show. We have seen some foreign-made refractometers show the burst point instead, so be sure to get one that measures the freeze point. You should also test the pH of the antifreeze. You use a pH meter for this test or use litmus paper.

The pH should never be below 7.5. Note that the lower the number, the higher the acid content, and 7 is neutral.

Solar Swimming Pool Systems

At the start of the heating season you should check the controller for proper operation. Visually inspect the collectors and piping. Make sure all the fittings are still connected. Clean any debris that may have accumulated around the collectors. Turn the system on and check for leaks. Turn the system off and make sure it drains properly.

Air Collector Systems

The weak links in air systems are the dampers. Even the best of them just do not last very long. The dampers should be checked and lubricated at least once a year at the start of the heating season. The blower should also be checked and lubricated at the start of every heating season.

One thing that can affect the performance of air collectors is dust. On all air systems except transpired systems, air is drawn from inside the building and circulated through the collector. If this air has any dust in it, that dust can build up on the absorber plate and significantly impede heat transfer and decrease the collector's efficiency. To minimize this, an air filter must be placed in the cold-air duct. This air filter must be checked periodically, especially where dusty conditions exist. Media filters with lots of surface area are the best filter choice and will last the longest between replacements.

For additional current information about solar thermal systems, please refer to Tom Lane's book *Lessons Learned 1977 to Today*, available from the author at ecssolar.com.

Appendix A
Wire Sizing Chart

12 Volt 2% Wire Loss Chart										
					Wire Size					
Amps	**#14**	**#12**	**#10**	**#8**	**#6**	**#4**	**#2**	**#1/0**	**#2/0**	**#4/0**
1	45	70	115	180	290	456	720			
2	22.5	35	57.5	90	145	228	360	580	720	1060
4	10	17.5	27.5	45	72.5	114	180	290	360	580
6	7.5	12	17.5	30	47.5	75	120	193	243	380
8	5.5	8.5	15	22.5	35.5	57	90	145	180	290
10	4.5	7	12	18	28.5	45.5	72.5	115	145	230
15	3	4.5	7	12	19	30	48	76.5	96	150
20	2	3.5	5.5	9	14.5	22.5	36	57.5	72.5	116
25	1.8	2.8	4.5	7	11.5	18	29	46	58	92
30	1.5	2.4	3.5	6	9.5	15	24	38.5	48.5	77
40			2.8	4.5	7	11.5	18	29	36	56
50			2.3	3.6	5.5	9	14.5	23	29	46
100					2.9	4.6	7.2	11.5	14.5	23
150							4.8	7.7	9.7	15
200							3.6	5.8	7.3	11
					Maximum Distance (feet)					

Appendix B
METRIC CONVERSIONS

To Convert	From Metric	To US	Multiply By
Temperature	C	F	9/5 + 32
Delta temperature	C	F	1.8
Length	meters	feet	3.28
Length	millimeters	inches	0.04
Area	square meters	square feet	10.76
Volume	cubic meters	cubic feet	35.31
Liquid volume	liters	gallons	.26
Storage Volume	liters	gallons	.02429
Collector area	square meters	square feet	.024
Speed	meters/second	mph	2.2369
Power	watts	btu/hour	3.41
Power	watts/square meter	watts/square foot	0.09
Pump power	watts	horsepower	0.0013
Pump power	kilowatts	horsepower	1.34
Heat exchanger power	kilowatts	btu/hour	3412
Solar resource	$kWh/m^2/day$	$btu/ft^2/day$	316.99

To Convert	From Metric	To US	Multiply By
Solar radiation	MWh/m^2	Million btu/ft^2	0.32
Energy	MWh	Million btu	3.41
Energy	liter	gallon	.26
Energy cost	$/MWh	$/Million btu	.29
Energy cost	$/liter	$/gallon	3.78
Energy cost	$/cubic meter	$/ccf	2.83

(ccf=$100ft^3$ = 1 therm)

Appendix C

Temperature Vs. Resistance Conversion Chart For 10K Thermistors													
°F	OHMS	°F	OHMS	°F	OHMS	°F	OHMS	°F	OHMS	°F	OHMS	°F	OHMS
-50	491,142	0	85,387	50	19,900	100	5,827	150	2,044	200	829	250	378
-49	472,642	1	82,719	51	19,377	101	5,697	151	2,005	201	815	251	373
-48	454,909	2	80,142	52	18,870	102	5,570	152	1,966	202	802	252	367
-47	437,907	3	77,656	53	18,377	103	5,446	153	1,929	203	788	253	362
-46	421,602	4	75,255	54	17,899	104	5,326	154	1,892	204	775	254	357
-45	405,965	5	72,937	55	17,435	105	5,208	155	1,856	205	763	255	352
-44	390,966	6	70,698	56	16,985	106	5,094	156	1,821	206	750	256	347
-43	376,577	7	68,535	57	16,548	107	4,982	157	1,787	207	738	257	342
-42	362,770	8	66,447	58	16,123	108	4,873	158	1,753	208	726	258	337
-41	349,522	9	64,428	59	15,711	109	4,767	159	1,720	209	714	259	332
-40	336,804	10	62,479	60	15,310	110	4,664	160	1,688	210	702	260	327
-39	324,597	11	60,595	61	14,921	111	4,563	161	1,657	211	691	261	323
-38	312,876	12	58,774	62	14,543	112	4,464	162	1,626	212	680	262	318
-37	301,622	13	57,014	63	14,176	113	4,368	163	1,596	213	669	263	314
-36	290,813	14	55,313	64	13,820	114	4,274	164	1,567	214	658	264	309
-35	280,433	15	53,669	65	13,473	115	4,183	165	1,538	215	648	265	305
-34	270,460	16	52,078	66	13,136	116	4,094	166	1509	216	637	266	301
-33	260,878	17	50,541	67	12,809	117	4,007	167	1,482	217	627	267	296

Temperature Vs. Resistance Conversion Chart For 10K Thermistors													
°F	OHMS	°F	OHMS	°F	OHMS	°F	OHMS	°F	OHMS	°F	OHMS	°F	OHMS
-32	251,670	18	49,054	68	12,491	118	3,922	168	1,455	218	617	268	292
-31	242,821	19	47,616	69	12,182	119	3,839	169	1,428	219	607	269	288
-30	234,316	20	46,225	70	11,882	120	3,758	170	1,402	220	598	270	284
-29	226,138	21	44,879	71	11,589	121	3,679	171	1,377	221	588	271	280
-28	218,276	22	43,577	72	11,305	122	3,602	172	1,352	222	579	272	276
-27	210,716	23	42,318	73	11,029	123	3,527	173	1,328	223	570	273	273
-26	203,445	24	41,099	74	10,761	124	3,454	174	1,304	224	561	274	269
-25	196,451	25	39,919	75	10,500	125	3,382	175	1,281	225	553	275	265
-24	189,722	26	38,777	76	10,246	126	3,312	176	1,258	226	544	276	262
-23	183,248	27	37,671	77	9,999	127	3,244	177	1,235	227	536	277	258
-22	177,019	28	36,601	78	9,758	128	3,177	178	1,213	228	527	278	255
-21	171,023	29	35,565	79	9,525	129	3,112	179	1,192	229	519	279	251
-20	165,251	30	34,561	80	9,297	130	3,049	180	1,171	230	511	280	248
-19	159,696	31	33,590	81	9,076	131	2,987	181	1,150	231	503	281	244
-18	154,347	32	32,648	82	8,861	132	2,926	182	1,130	232	496	282	241
-17	149,197	33	31,737	83	8,651	133	2,867	183	1,110	233	488	283	238
-16	144,236	34	30,853	84	8,447	134	2,809	184	1,091	234	481	284	235
-15	139,458	35	29,998	85	8,249	135	2,752	185	1,072	235	473	285	232
-14	134,855	36	29,169	86	8,056	136	2,697	186	1,054	236	466	286	229
-13	130,420	37	28,365	87	7,867	137	2,643	187	1,035	237	459	287	225
-12	126,147	38	27,587	88	7,684	138	2,591	188	1,017	238	452	288	223
-11	122,030	39	26,832	89	7,506	139	2,539	189	1,000	239	445	289	220
-10	118,061	40	26,100	90	7,333	140	2,489	190	983	240	439	290	217
-9	114,235	41	25,391	91	7,164	141	2,440	191	966	241	432	291	214
-8	110,547	42	24,704	92	6,999	142	2,392	192	950	242	426	292	211
-7	106,991	43	24,037	93	6,839	143	2,345	193	933	243	420	293	208
-6	103,561	44	23,391	94	6,683	144	2,299	194	918	244	413	294	206
-5	100,254	45	22,764	95	6,530	145	2,254	195	902	245	407	295	203
-4	97,063	46	22,156	96	6,382	146	2,210	196	887	246	401	296	200
-3	93,986	47	21,566	97	6,238	147	2,167	197	872	247	395	297	198
-2	91,017	48	20,993	98	6,097	148	2,125	198	857	248	390	298	195
-1	88,152	49	20,438	99	5,960	149	2,084	199	843	249	384	299	193
												300	190

Glossary

Absorber: The part of a solar collector that absorbs the sun's energy and changes that energy into heat.

Absorptance: The efficiency of an absorber. A ratio of the amount of energy hitting a surface to the amount of energy that the surface actually absorbs.

Absorptive coating: A coating for absorber plates that maximizes the plate's absorptance.

Active solar energy system: Any solar energy system that uses a collector to absorb solar energy.

Ambient temperature: The temperature of the surrounding air.

Aquastat: A thermostat that measures the temperature of a liquid.

Azimuth: The distance in degrees from north. (South would have an azimuth of 180).

Ball valve: A valve that uses a ball with a hole through its center to control the flow of liquid through a piping circuit.

Batch heater: Another name for ICS collector.

Battery: A storage device for energy.

Boiler control: A hydronic heating system component that includes a low-voltage power supply and a heavy-duty relay for switching purposes.

British Thermal Unit (Btu): The amount of energy it takes to raise the temperature of one pound of water one degree F.

Centrifugal pump: A pump that moves a fluid by spinning it with enough force to throw the fluid outward.

Check valve: A valve that allows a fluid to travel in only one direction within a circuit.

Circulator: A pump designed to move a fluid within a circuit.

Closed-loop solar heating system: A type of system where the solar fluid stays in the solar loop and does not mix with the domestic water.

Collector: A device that collects solar energy.

Collector loop: Another name for solar loop.

Collector tilt: The angle between the solar collector plane and the horizontal plane.

Conduction: Heat flow within a material that is caused by the difference of temperature within that material.

Convection: The movement of parts of a fluid because of variations in the fluid's density caused by temperature differences.

Counterflow heat exchanger: A heat exchanger where the two fluids pass each other in opposite directions.

Coupler: A pipe fitting that joins two pieces of pipe together.

Degree day: The difference between 65°F and the ambient temperature.

DHW: Domestic hot water.

Differential temperature controller: An electrical device that measures the difference in temperature between two locations and switches a pump on or off in relation to the difference in the temperatures at the two locations.

Diffuse radiation: Solar radiation that is scattered by the atmosphere and everything that is suspended in the atmosphere, particularly water vapor.

Direct radiation: Solar radiation that has not been scattered.

Double-walled heat exchanger: A liquid-to-liquid heat exchanger whose separation wall between the two fluids is made of two layers.

Drainback: A solar water heating method where the solar fluid is pumped to the collectors and fills the solar-loop piping when solar energy is available for harvest but drains back to a holding tank when not collecting energy.

Draindown system: A solar water heating system that drains when not in use.

Emittance: The property of a material to radiate energy.

Expansion tank: A tank used in solar energy systems to compensate for the expansion of liquids in a closed-loop circuit.

Evacuated tube collector: A collector that uses absorber plates that are enclosed in a glass tube that has a vacuum inside of it.

Flat plate collector: A rectangular solar thermal collector, typically four inches deep, four feet wide and eight feet or ten feet tall.

Flow meter: A device that measures the flow of a liquid within a pipe. It is often a clear pipe with a scale on the side.

Flow rate: A measure of the rate of a certain volume of fluid flowing through a circuit.

Fossil fuel: A carbon-rich and energy-dense fuel that was created from concentrated organic matter that lived or grew on the Earth millions of years ago.

Friction: The resistance to movement created when two materials rub against each other.

Friction head: The pressure created in a circuit resulting from a fluid moving through a pipe.

Glazing: A transparent and weatherproof covering.

Head: The total pressure or resistance to the flow of a liquid within a circuit exerted by gravity and friction.

Header: A manifold.

Heat exchanger: A device that facilitates the transfer of heat from one fluid to another.

Heat transfer fluid: A fluid that is used to transfer heat from one location to another. This fluid could be a liquid or a gas. Also called solar fluid.

Hydronic: The use of a liquid to transfer heat.

Indirect system: A solar water heating system where the domestic water is not heated in the collectors but is heated using a heat transfer fluid and a heat exchanger.

Impellers: The vanes in a pump that spin and move the fluid.

Insolation: The total amount of solar radiation hitting a surface.

Liquid collector: A collector that uses a liquid as the heat transfer fluid.

Manifold: A pipe with several outlets.

Open-loop system: A kind of solar water heating system in which some part of the system is vented or the solar loop contains potable water.

Orientation: The direction the solar collectors face in relation to south.

Passive systems: Solar heating systems that do not use a collector to absorb the sun's energy. A term used to describe a solar heating system where the house itself is the solar collector.

Photovoltaic panel: A type of solar collector that absorbs solar energy and converts it to electrical energy. Also called PV.

Potable water: Drinkable water.

Propylene glycol: A nontoxic antifreeze used in solar heating systems. It is mixed with water and used as a heat transfer fluid.

Pump: A device that moves a fluid through a circuit.

Pump curve: A chart that shows the performance of a pump over varying conditions.

Radiation: The flow of energy through a space.

Renewable energy: Energy that is replenished by nature on a regular basis.

Risers: The pipes that connect the top and bottom manifolds in an absorber plate.

Selective surface: A surface that absorbs solar energy very efficiently but does not radiate solar energy very well.

Sensor: A device that identifies the temperature at a certain location.

Solar collector: A device that collects the energy of the sun that is radiated to the Earth's surface.

Solar energy: Energy produced by the Sun that is radiated to the Earth's surface.

Solar fluid: The heat transfer fluid used in a solar heating system.

Solar loop: The circuit of piping that travels from the collectors to the heat exchanger/storage and back to the collectors.

Specific heat: The amount of heat, in Btu, needed to raise the temperature of one pound of a substance by one degree F.

Stagnation: When the solar fluid in a closed-loop solar heating system is not circulating.

Static head: The pressure created by a column of water in a pipe.

Therm: 100,000 Btu.

Thermosiphon: The movement of a fluid caused by convection.

Union: A pipe fitting that joins two pipes together and is capable of coming apart.

Volute: The part of a pump that houses the impellers. Also called a pump head.

Wrench: A solar installer.

INDEX

B

C

D

About the Authors

Bob Ramlow

Bob started in the renewable energy business in 1976 and has been involved in renewable energy system installations, sales and manufacturing. He has taught workshops about solar energy throughout the US. He is an ISPQ Certified Solar Thermal Independent Master Trainer, a NABCEP™ Certified Solar Thermal Installer and a founding member of the Midwest Renewable Energy Association. He continues to teach and write about solar energy from his home in Wisconsin.

Benjamin Nusz

Benjamin Nusz has worked as an installer, site assessor, designer and consultant for solar thermal systems. He is a NABCEP™ Certified Solar Thermal Installer and is currently training the next generation of solar professionals as the Renewable Thermal Instructor at Mid-State Technical College in Wisconsin.

If you have enjoyed *Solar Water Heating* you might also enjoy other

BOOKS TO BUILD A NEW SOCIETY

Our books provide positive solutions for people who want to make a difference. We specialize in:

Sustainable Living • Green Building • Peak Oil • Renewable Energy
Environment & Economy • Natural Building & Appropriate Technology
Progressive Leadership • Resistance and Community
Educational and Parenting Resources

New Society Publishers

ENVIRONMENTAL BENEFITS STATEMENT

New Society Publishers has chosen to produce this book on Enviro 100, recycled paper made with **100% post consumer waste**, processed chlorine free, and old growth free.

For every 5,000 books printed, New Society saves the following resources:[1]

[1]Environmental benefits are calculated based on research done by the Environmental Defense Fund and other members of the Paper Task Force who study the environmental impacts of the paper industry.

For a full list of NSP's titles, please call 1-800-567-6772 *or check out our website at:*

www.newsociety.com

NEW SOCIETY PUBLISHERS